Dear Engineering Student,

You are very important to me. You are obviously important to yourself and definitely important to all of society. You are preparing for an incredibly important profession.

Now is a time of unprecedented opportunity for you as an engineering student to develop yourself for a career that optimizes your best skills, interests, values, and societal need. Aim for that place of opportunity and, with desire and perseverance, you'll reach it. There, in the right environment, you will solve important problems, perhaps create new technology, and enjoy a career of lasting reward and significance.

As you'll see in this book, I am passionate about showing how you can prepare to fulfill your dreams and be the best that you want to be, professionally and in life.

You'll want to study and be technically competent—that's the foundation for your engineering career. But that "best that you want to be" goal extends beyond your technical skills. We are talking about using your total potential by applying your combined technical, professional, and personal skills. It means selecting a major and then a career that is best for you—based on your talent and where you choose to contribute.

Quoting President Theodore Roosevelt, "far and away the best prize that life has to offer is the chance to work hard at work worth doing." That timeless truth is as relevant today as it was when Teddy Roosevelt spoke it. One hundred years later Steve Jobs, founder and CEO of Apple Inc., says, "I want to put a ding in the universe . . . I think we're having fun. I think our customers really like our products. And we are always trying to do better." As an engineer, following President Roosevelt's and Steve Jobs' thinking can lead you to a meaningful life, a rewarding career, and more than sufficient wealth.

I believe that dedication to your technical coursework combined with a balanced approach to your career development will give you the total range of education and skills that will allow you to have the right job or graduate school admission upon commencement. Moreover, you will create a future that is best for you. That goal is very important for you, your family, your country, and your planet.

I wish you well as you read this book and take action toward your successful future.

Sincerely,

Dean Millar

Ready for Takeoff!

A Winning Process for Launching Your Engineering Career

Dean C. Millar
School of Engineering
University at Buffalo

Prentice Hall
Upper Saddle River Boston Columbus San Francisco New York Indianapolis
London Toronto Sydney Singapore Tokyo Montreal Dubai Madrid
Hong Kong Mexico City Munich Paris Amsterdam Cape Town

Vice President and Editorial Director, ECS: Marcia J. Horton
Executive Editor: Holly Stark
Editorial Assistant: Keri Rand
Vice President, Production: Vince O'Brien
Senior Managing Editor: Scott Disanno
Production Liaison: Jane Bonnell
Production Editor: Shiny Rajesh, Integra
Senior Operations Supervisor: Alan Fischer
Operations Specialist: Lisa McDowell
Executive Marketing Manager: Tim Galligan
Marketing Assistant: Mack Patterson
Art Director and Cover Design Director: Jayne Conte
Cover Designer: Bruce Kenselaar
Cover Image: Getty Images Inc. - Stone Allstock
Art Editor: Greg Dulles
Media Editor: Daniel Sandin
Composition/ Full-Service Project Management: Integra Software Services Pvt. Ltd.

Library of Congress Cataloging-in-Publication Data
Millar, Dean C.
Ready for takeoff! : a winning process for launching your engineering career / Dean C. Millar.—1st ed.
p. cm.
Includes bibliographical references.
ISBN-13: 978-0-13-608127-2 (alk. paper)
ISBN-10: 0-13-608127-4 (alk. paper)
1. Engineering—Vocational guidance. I. Title.
TA157.M489 2011
620.0023—dc22
2 3 4 5 6 7 8 9 CRS 15 14 13 12 11 10 2010028201

Prentice Hall
is an imprint of

www.pearsonhighered.com

ISBN-13: 978-0-13-608127-2
ISBN-10: 0-13-608127-4

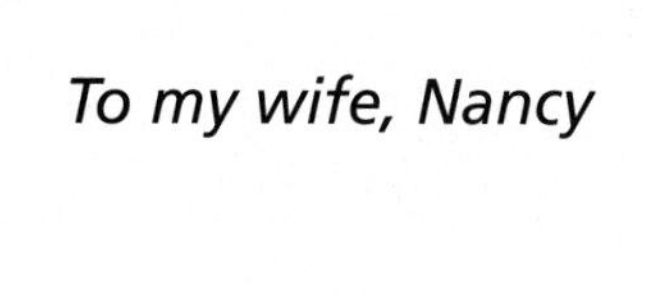

To my wife, Nancy

PREFACE

Ready for Takeoff! is a "how to" text, written specifically to assist you in launching your career. The book is based on a culmination of the author's 28 years of industrial experience along with 17 years delivering a career preparatory course for engineering students. The material in this book is a vitally important component of an undergraduate engineering education at any university. It supplements essential technical coursework with logical steps for career development. The book also includes advice from engineering professionals and university experts on a broad range of subjects that are important for on-the-job success.

Ready for Takeoff! can be used as a career course text or as a stand-alone career success handbook. It is designed for engineering students and instructors who want to complement their technical curriculum with complete professional preparation. This includes comprehensive engineering career planning, effective job search, and professional success subjects for transition from school to ready-to-start employment.

Engineering undergraduate students are well prepared with engineering theory and fundamentals when they graduate but generally lack broader professional success skills. See Chapter 2, pages 16–19 for evidence. This imbalance diminishes the likelihood that undergraduate students will have sufficient career development tools for the right job offer or graduate school decision by commencement. A right job offer or grad school decision is one that students find is the best planned use of their strengths, interests, and values. *Ready for Takeoff!* empowers students to aim a self-assessment and marketing plan to achieve employment in an individually correct market niche for their career success and life happiness.

Most engineers will go into the engineering profession; however, opportunities for an engineering graduate are not limited to doing traditional engineering work. During the 21st century, we see an engineering degree to be an educational foundation for many functions and professions. An engineer can move into management of almost any enterprise. Increasingly, presidents of companies will have earned an engineering undergraduate degree, perhaps with an MBA. Education, medicine, law, and finance are all professions that benefit from employing engineering degree graduates. Where can you release your most genuine passion and abilities? *Ready for Takeoff!* poses that question and will help students get there and succeed.

"Part 1: Pre-Employment Success: Understanding Yourself and Getting the Right Job" explains step-by-step actions for engineering undergraduates to accomplish personal assessment and job search skills, obtain career related co-op/internship employment, and secure the best-for-them job offer when they graduate at the bachelor's, master's, or doctoral level. It includes a four- or five-year career development plan, with action steps for each undergraduate year.

"Part 2: Professional Functions and Opportunities" and **"Part 3: Personal and Professional Success Skills"** contain a unique set of chapters. Together, they address essential engineering professional success subjects, offering wisdom from 32 appropriate industrial and university experts, enabling engineering students to "hit the ground running" when they take off to the job.

The chapter topics have been carefully chosen and have evolved as a collection of class subjects over a 17-year period. The choice of topics and their content are continuously validated by engineering students and their industrial mentors, as well as by the volunteer seminar presenters and the author. The presenters comprise a powerful means of exposing the students to leaders in their fields, many of whom are among their school's most successful alumni.

ABET, Inc. (formerly the Accreditation Board for Engineering and Technology) requires proven technical as well as broader educational outcomes in their *Criteria for Accrediting Engineering Programs*. ***Ready for Takeoff!*** **addresses ABET's "Criterion 3, Program Outcomes,"** listed below by ABET nomenclature followed by the associated *Ready for Takeoff!* chapters.

- **(a)** "an ability to apply knowledge of mathematics, science, and engineering": Chapters 11–26
- **(d)** "an ability to function on multidisciplinary teams": Chapters 21–24
- **(f)** "an understanding of professional and ethical responsibility": Chapter 16
- **(h)** "the broad education necessary to understand the impact of engineering solutions in a global, economic, environmental, and societal context": Chapters 11, 16, 18, 19, 24
- **(i)** "a recognition of the need for, and an ability to engage in life-long learning": Chapters 11, 18, 19, 25, 26
- **(j)** "a knowledge of contemporary issues": Chapters 4, 7, 10–19, 24
- **(k)** "an ability to use the techniques, skills, and modern engineering tools necessary for engineering practice": Chapters 1–26

In summary, *Ready for Takeoff!* challenges engineering students to create their own future and gives them the information and encouragement to do it. It's a professional and life success book that will help those who read and act on its contents to take off, well prepared, into the engineering careers that suit them best.

ACKNOWLEDGMENTS

Many people made invaluable contributions to this book. Initial credit goes to Dr. George Lee, former Dean of Engineering at the University at Buffalo, State University of New York. In 1994, Dr. Lee had the visionary commitment to begin an Engineering Career Institute. His desire was to supplement engineering students' strong technical education with broader professional skills and job experience that would significantly impact their career success, including the probability of a job offer at graduation.

Fortunately, the Engineering Career Institute has been successful for 17 years, due in large measure to the enthusiastic participation of engineering students as well as the strong support of faculty and administration. In addition to Dr. Lee, I especially want to thank former Dean Mark Karwan and current Dean Harvey Stenger for their steadfast support of this program. Special appreciation goes to my immediate supervisor, Senior Associate Dean Robert Barnes, who has given me guidance and encouragement throughout my time at UB.

This book would not be possible without the great wisdom and experience of our 32 engineering professional contributors featured in Parts 2 and 3. Their contributions allow engineering students to receive the "full package" of engineering professional skills envisioned by Dr. Lee. Sincere thanks to you all. You are contributing significantly to the professional success of engineering students.

Suggestions from reviewers at major universities throughout the country were extremely helpful and appreciated. Specifically, I thank Helen Oloroso at Northwestern University, Mark Savage at Cornell University, Jack Rayman at Penn State University, Andrew Duffy at Drexel University, Anita Todd at University of Cincinnati, Thomas Akins at Georgia Tech, Craig Gunn at Michigan State University, Fred Barez at San Jose State University, and Louise Carrese at the Rochester Institute of Technology.

The editorial staff at Pearson Education has been superb in producing this book. I thank, in particular, Executive Editor Holly Stark, Senior Managing Editor Scott Disanno, Project Manager Jane Bonnell, Copy Editor Karen Ettinger, and Editorial Assistant Keri Rand. Also, thank you to Field Editor Sean Wittmann, who provided suggestions and optimism at the outset.

Special thanks to two good friends and communication experts who gave valuable editing assistance. Bill Grunert contributed Chapter 25, "Effective Writing and Presentation Skills." He and Tom Holub provided great editorial advice.

Why should I thank my immediate family last? Perhaps to leave a lasting impression of my love and gratitude for them. My wife and best friend Nancy encouraged me to write this book at an anniversary dinner. She has been rock-solid in the support, sacrifice, and editorial advice that go with being an author's wife. Our daughter Jenn and son Greg provided editing assistance and invaluable support to their parents throughout the book's creation.

And to you, engineering student: best wishes as a future engineer whose career success is the reason for this book.

ABOUT THE AUTHOR

Dean Millar's passion for personal and professional development began when he served as a training and education officer in the U.S. Navy. After receiving an M.S. degree emphasizing human resources, the author enjoyed 28 years with Union Carbide/Praxair Inc. in human resources management. Millar's industrial career was spent in an engineering environment including research, development, design and project engineering, manufacturing, computer technology, and operations management. With job titles including Technical Personnel Manager and Human Resources Manager, Millar offers substantial experience as a leader and individual contributor in engineering recruitment, training, career development, college relations, and career transition. He led or administered teams focused on the development of personal, professional, and management skills of an engineering workforce. In addition, he was a founder and director of Buffalo-area Engineering Awareness for Minorities (BEAM), a minority youth engineering educational preparation program.

Ready for Takeoff! originated in 1994, when Dr. George Lee, then Dean of Engineering at the University at Buffalo, offered Millar a position of Assistant Dean to begin a unique Engineering Career Institute. Dean Lee wanted the author to use his industrial experience to create a program to complement engineering coursework to give engineering students real-world professional skills, including pre-employment classes and credit-worthy industrial employment experience. The goal was to enhance engineering students' career success, including a greater probability of a job offer at graduation, whether at the bachelor's, master's, or doctoral level. Subsequently, Millar helped initiate and has directed the Engineering Co-Op Program at UB.

For the past 17 years Millar has helped over 2000 engineering students launch their careers through the Engineering Career Institute, Internships, and Co-Op Programs. He has been rewarded by positive feedback from students and employers and by observing their winning relationships with each other and the university. Millar has received recognition for the success of these programs including a State University of New York Chancellors Award for Excellence in Professional Service and awards for Positive Influence on Students who were surveyed by UB one year after graduation.

Beyond work, Millar likes to spend time with his wife skiing, sailing, kayaking, and traveling. He also plays tenor saxophone with the Abino Bay Stompers, a Dixieland jazz band.

CONTENTS

Pre-Employment Success: Understanding Yourself and Getting the Right Job

PART 1

CHAPTER 1

Introduction and Your Four-Year Plan of Career Development

WHY IS THIS BOOK IMPORTANT TO YOU?

Ready for Takeoff! addresses significant engineering educational needs:

- ❖ A unique career development text to enhance your engineering education
- ❖ A guide for you, the student, to help you launch a successful engineering career
- ❖ A handbook for key subject reference during your engineering career

In a nutshell, this book supplements your engineering coursework to provide "what," "why," and "how" information for you to successfully land the right job and jump-start your career when you graduate, whether at the bachelor's, master's, or doctoral level.

WHAT DOES IT TAKE FOR YOU TO BECOME THE IDEAL ENGINEERING EMPLOYMENT CANDIDATE?

Four Keys to Your Success

1. A SOLID ENGINEERING EDUCATION

You need to have a sound grasp of engineering theory and fundamentals in your major. Your diligent study in an accredited college engineering curriculum will give you the theoretical, applied lab, and computer knowledge to meet the technical requirements to be an entry-level engineer.

2. CO-OP/INTERNSHIP EXPERIENCE

Before you graduate, you should have some career-related, on-the-job experience where you can apply your theoretical knowledge to achieve tangible project results. Co-op and internship programs provide great value. You will:

- ❖ Gain exposure to a real engineering employment situation
- ❖ Demonstrate technical and teamwork skills

- ❖ Obtain applied relevance of your engineering curriculum
- ❖ Use engineering principles to achieve project results
- ❖ Build evidence for your resume that you have created value
- ❖ Learn non-technical business skills vital to career achievement
- ❖ Assess, mutually with your employer, your "fit" for a career with them
- ❖ Increase your odds of a permanent job with your co-op or internship employer

3. CAREER SUCCESS SKILLS

This text supplements your engineering coursework and guides you toward the right job offer and career success by providing three parts to the book.

Part 1: Pre-Employment Success. Chapters 1–10 focus on the action steps from now until graduation that will help you get the right job. That includes self-analysis, your resume, use of your career center, interviewing, and executing a successful job search campaign. The job search includes where to look, how to look, and how to present yourself as the best qualified candidate for the right job.

Part 2: Professional Functions and Opportunities. Chapters 11–19 feature experts from industry and academia to explain important functions and success tips for you while in graduate school and then for industrial and academic careers. Each chapter provides important knowledge for you to apply in your engineering profession. There is generally not enough room in the undergraduate curriculum for these subjects. Yet they are all extremely important when you are on the job, whether in industry, government, academia, or as an entrepreneur. Awareness of these professional functions and opportunities will give you a competitive edge to take the right job and achieve early success.

Part 3: Personal and Professional Success Skills. Chapters 20–26 give you information from experts that can help you be very successful in your engineering professional career. Those skills include self-reliance, teamwork, leadership, communication, and effective transition from school to your profession. These are non-technical skills but crucial for your success, whether you are in a technical or managerial capacity.

4. A FOUR-YEAR CAREER DEVELOPMENT PLAN

First-year engineering students need incentive and hope for their professional future, amidst the rigors of calculus and physics. In fact, throughout all four undergraduate years you, as an engineering student, will profit from a plan for each year.

The following four-year engineering undergraduate process with suggested academic and career steps for each year was developed from the author's experience as well as a variation of several academic and career development models.

Co-op universities often have a five-year academic program. While a five-year curriculum does not exactly parallel a four-year program, these career development steps remain the same, and can be used at appropriate times during either four or five years.

SUGGESTED FOUR-YEAR UNDERGRADUATE SUCCESS PLAN

First Year

Get off to a good start.

- *Set academic goals.* Setting positive targets for your grades will help you achieve them. Keep up with the class. Get help from your professors during office hours, teaching assistants, or tutors as needed. Grades are important; they quantify how much you have learned in order to be a competent engineering professional. Employers and graduate schools look closely at your grades before hiring or admission. Knowing that, you don't want to limit your future opportunities due to grades lower than you are capable of achieving. So use your initiative and campus resources to form early academic success habits.
- *Develop self-reliance, planning, and time management skills.* These keys to your success in college and in your career are covered in Chapter 20. Make self-reliance, planning, and time management positive habits, starting now.
- *Embrace physics, calculus, and other first-year required courses.* These courses provide necessary fundamental knowledge for your major. You are intellectually building toward your major and your engineering career.
- *Explore engineering majors.* Your major should combine your greatest skills, interests, and professional target areas. Discuss options with an academic advisor. Talk with professors and students in your possible majors for more information. Attend department events and seek information to develop network connections and awareness of opportunities.
- *Explore foreign language course options.* Do this soon, especially if you are interested in international work—there is increasing opportunity worldwide.
- *Get involved.* Keep up with your studies and manage your time to explore engineering clubs or other activities. Develop your communication, teamwork, and leadership skills along with your engineering coursework. Start a portfolio to document your involvement on projects or in organizations. Keep evidence of your accomplishments in and out of the classroom as employers will focus on them.
- *Use your career services office.* Here you can learn more about careers in majors that are in line with your abilities and interests. Use the many career office resources, including resume assistance, job opportunity Web site, career library, career workshops, and technical job fairs. Chapter 6 describes career centers in greater detail.

❖ ***Take interviews for summer employment.*** A summer job provides experience and money. If possible, find a job with an employer related to your possible major. Test your skill and interest in that occupation. Learn how to succeed on a project and work with people. Establish and keep evidence of your accomplishments. Quantitative evidence is best; a result followed by how much money you saved or generated, expressed in dollars or percentages. Your supervisor should agree with evidence of your results—the company will appear on your resume and your supervisor might be an employment reference. Chapters 2 through 10 describe the pre-employment process in greater detail.

Second Year

Assess yourself, obtain career information, choose your major, and find an internship.

❖ ***Do a self-assessment of your skills, interests, goals, values, and opportunities leading to a choice of a major and possible career interest.*** Learn from the results of your activities during your first year. What are you good at? What is your passion? What do you want to do? Chapter 3 is devoted to self-assessment. You can also receive help on self-assessment from your career center.

❖ ***Complete general courses leading to your major.*** Consult your faculty advisor early in the fall semester to ensure you are taking the right prerequisite courses for your intended major.

❖ ***Continue to use your career services office.*** Prepare or update your resume/cover letter and upload it onto the career services Web site. Take interviews for co-op, summer, or part-time employment related to your objectives. Take advantage of the range of resources at your career services office, including career preparation workshops and career fairs. Some of these services are not only through your career center. At your university, they may be offered by the division of professional practice or your co-op program office. Attending a career fair early is important, since it exposes you to a wide range of employers and career options. Attend career fairs no later than your second year, and each year thereafter. That will help you target your goals and give you an opportunity to line up summer or part-time employment related to your objectives.

❖ ***Network through information interviews.*** Confirm your choice of major by talking with professionals who have jobs related to your intended major. Build your network through faculty, career services, professional clubs, alumni, family, friends, and other students. Use online technology such as LinkedIn® and Facebook as a resource. Besides career information and contact tips, your network may produce a summer job opportunity suited to your goals. Networking is extremely important for many purposes, and it's good to take early advantage of this. Chapter 7 describes job searching in greater detail.

- ❖ ***Balance coursework with campus activities.*** Join a student club or organization. Continue to build communication, teamwork, and leadership skills.
- ❖ ***Prepare for a co-op or internship opportunity.*** Apply your technical coursework on a job that parallels your abilities and interests. Chapters 4, 10, 22, and 26 describe this in greater detail.
- ❖ ***Declare your major, and possibly minor, by the end of your second year.*** This decision will be made following your self-assessment and your investigation of careers that optimize your strengths and interests.
- ❖ ***Explore participation and plan for an international education experience, research, work, or service learning.*** Plan now for your participation during your third or fourth year. Chapter 19 describes this in greater detail.
- ❖ ***Get co-op or internship experience during or after your second year.*** Ideally, your job will relate to your major and tentative career goals. Aim for a mixture of applied technology and teamwork. If you're thinking of graduate school, a research project could be good for you. Chapters 3–10, 18, 22, and 26 describe this in greater detail.

Third Year

Dive into your major, be active in a professional organization, and get co-op/internship experience.

- ❖ ***Plan with your academic advisor*** to ensure you will meet requirements for your major and graduation.
- ❖ ***Take courses in your major*** while assessing that your major best suits your abilities, interests, and objectives.
- ❖ ***Order business cards through your university print shop.*** The cards could include the name of your school and an attractive logo. Include your name, degree and year/month of graduation, address, phone number, and e-mail address. Exchange business cards during information interviews and other networking opportunities.
- ❖ ***Explore employment opportunities and graduate school.*** Network with professionals in your area of interest. Speak with faculty about possible specialization in graduate school. Evaluate options that will be available after you graduate. Grad school? Employment? Pursue both at once with results pending? Make a plan to implement your employment and graduate school application process, beginning immediately in your senior year. If you have your eye on graduate school, during your junior year get to know three professors who can give you guidance and serve as references. These professors could be references for employment as well as graduate school. Chapters 2–10 describe employment in greater detail. Chapter 18 describes academic careers and graduate school in greater detail.
- ❖ ***Use your career services office, co-op office, or division of professional practice.*** Update your resume and cover letter and post it on your

school's job listing Web site. Attend a technical career fair in the fall to meet with employers for co-op or internship job opportunities in your major area of interest. Do practice interviews with a career coach. Participate in on-campus interviewing for a co-op or internship assignment. Continue to explore employment opportunities through the career services Web site and library.

- ***Be active in a student professional organization.*** Assume a leadership position and develop planning, organizing, and teamwork skills. Keep track of results to list on your resume and portfolio.
- ***Enroll in a co-op or internship course.*** Take full advantage of co-op or internship employment opportunities related to your major and employment objectives during or following your third year. Apply your major coursework to achieve quantifiable project results. Continue to develop teamwork and communication skills. Know that you and your employer are mutually assessing the possibility of a job offer when you graduate.
- ***Take the Graduate Record Exam (GRE)*** during the summer after your third year. You could precede the exam with a GRE preparation course. If you're thinking of graduate school, the GRE is generally a prerequisite.

Fourth Year

Get ready for takeoff—for graduation and for graduate school or employment. Take the following steps and you will maximize your options.

- ***Manage job offers and graduate school acceptances so that they come at the same time.*** You might think this is beyond your control, but proper planning can make it happen.
- ***Have your resume and cover letter ready early.*** Your resume and cover letter should be up to date, critiqued, and online with career services in early September. Include courses, grades, and results from your most recent co-op/internship experience.
- ***Be interviewed through on-campus recruiting*** with employers of interest, beginning in September. Campus recruiting starts earlier than many students think—early action yields best results. Chapter 6 describes this in greater detail. Continue to use the range of career services offerings. Some fourth-year students delay using the career services office until later in the year, putting them at a big employment disadvantage.
- ***Meanwhile, complete graduate school applications early.*** This looks like a lot of early action, and it is—but it is for your benefit to be able to compare employment opportunities and graduate school acceptances at the same time. In order to do that, an early start on both is essential. Consult with one to three professors who know you and who specialize in your technical area of interest. Those professors can recommend

graduate schools for you and also serve as references. Plan to obtain three faculty recommendations. Find the costs of applying to your graduate schools of interest, and budget for them. Chapter 18 describes this in greater detail.

- ❖ ***Order business cards*** with up-to-date contact information.
- ❖ ***Attend a technical career fair,*** often held in September or October. This fair will include employers seeking new graduates as well as, possibly, graduate school representatives. Be ready with your resume, advance knowledge of the companies attending, and a 20-second commercial that summarizes how you and your skills match the employers' needs. Most students don't do this preparation, so you will stand out and more likely be offered an interview after the fair.
- ❖ ***Check with your faculty advisor/undergraduate services group*** to ensure you will meet all requirements for graduation. Apply for graduation, according to your school's schedule.
- ❖ ***Connect with your former co-op/intern employer*** or an employer of interest to you. If possible, have a contact person in an organization who will personally champion your employment consideration. A supportive previous mentor is a good bet to arrange a personal discussion with a hiring manager in that organization.
- ❖ ***Visit employers and/or graduate schools*** to obtain offers and decide on acceptance, based on your options and goals. The ideal situation would be to have offers of employment and graduate school acceptance come at the same time. That's challenging to arrange, since you're not the one giving the offers. But start early and make your applications in parallel—that increases the odds of getting responses at the same time. Also, you can inquire about anticipated response time from both employers and universities, which will enable you to better manage that issue. Chapters 9 and 18 describe this in greater detail.
- ❖ ***Accept an employment or graduate school offer*** and notify your career services office and engineering college of your post graduation plans.
- ❖ ***Take the Fundamentals of Engineering (FE) exam*** on your way to becoming a Professional Engineer (PE). The FE exam is offered in April and October of each year. You can take it before or after you graduate. Chapter 16 describes engineering professionalism in greater detail.

ACTION SUMMARY

As a successful engineering student, select and excel in a major that suits your skills and interests. Meanwhile, throughout your time in school, take academic and career development action toward a rewarding engineering profession. Aim for a career that best suits your strengths, passions, and aspirations.

To reach your aspirations it is important that:

- ❖ You, the student, be thinking and taking action on your career future during all four undergraduate years. Ultimately, you are responsible for your future. The following chapters are designed with information to help you succeed over the next four or five years to attain your engineering professional goals.

❖ Engineering faculty, career services, and co-op administrators are available to provide you with resources to achieve optimal educational and career results. The goal is supporting your efforts to land a great job when you graduate and to enter the engineering profession with optimum preparation for success.

With your talent and multiple resources, you can reach your goals and successfully enter the important and honorable engineering profession. This book, with advice from many experts, will help you along the way.

Chapter Summary: Takeoff Tips

Ready for Takeoff! addresses important engineering educational needs:

❖ A unique career development text to enhance your engineering education
❖ A guide for you to launch a successful engineering career
❖ A handbook for key subject reference during your engineering career

To be the ideal engineering employment candidate, you need:

❖ A solid engineering education
❖ Co-op/internship experience
❖ Career success skills, as described in this book
❖ A four-year career development plan (five years for co-op schools)

Exercises

1. What are the four elements for you to be the ideal engineering employment candidate? Why are they each important?
2. Prepare a plan for each academic year and summer to advance your career development.

Looking Ahead

Chapter 2 will center on three engineering professional "life realities." These realities are important especially because they are often over-looked yet are fundamental to your professional and life success. Living the three essential realities will make you stand out as being "ready for takeoff."

Co-op Success Stories

Valuable Results from Student Employees

Here is some actual feedback from employers about the value that co-op students contributed to their organizations. Establishing a dollar or percentage value to co-op or internship results is great for you to put on your future resume to show your "track record" evidence of quantifiable success.

Northrop Grumman
Eric used Visual C++ to automate laboratory test procedures on new component devices in order to speed up the production of these devices. As a result of the improved test procedures, time spent on testing has decreased by 68%.

BOC Edwards/Precision & Vacuum Parts
Hilbert came up with a very good substitute vane material for vacuum pumps. This reduced the cost of these vanes by over 90%.

BMP America
Tomasz's analysis of our oil roller production process allowed him to incorporate several design changes that increased productivity. He reorganized the material database into an improved and user-friendly format. Both projects were immediately implemented and each has displayed successful results. The estimated value of his work results in savings of $40,000 per year.

United Rentals
Shaun, with an employee teammate, designed an automatic device that reduced production cycle time by 80%.

Goodyear Dunlop Tires
Jennifer developed part of an improved steam line program that is being implemented. The result will be a projected 20% cost savings of $500,000/yr.

HBE Company
Maureen was responsible for changes in sub-contracting and purchase orders that totaled $134,350. She also saved the company about $8,000 by negotiating quotes with the contractors.

Reichert
Bob designed a new test protocol for a laboratory instrument that was about to be released by the company. The protocol was instituted as a standard procedure for all new instruments.

Motorola
Chris saved a minimum of $50,000 by building a more economical way of manufacturing heated seat modules. He accomplished this by narrowing process flow options and obtaining budgetary cost estimates from equipment suppliers.

Praxair

Kathryn recovered $100,000 in capital investment by troubleshooting and developing methodologies on a laboratory analytical instrument.

Prestolite Electric

Without the competent assistance of our UB engineering interns, we may not have accomplished our goal of QS 9000 certification. They worked on elements of product flow, operator's instructions, and other advanced product quality planning requirements for QS 9000.

Xerox Corporation: Ink Jet Focus Factory

Sawson demonstrated good problem-solving skills and saved the company $20,000.

CHAPTER 2

Important Life Realities

You know that engineering study is not easy. As an engineering student, you need to be focused on the technical subject matter of your major. Because of the demands of an engineering curriculum, you may not have had much opportunity to pause and consider some of the other factors of your future life and career direction.

The three realities in this chapter are extremely important, yet often tend to be overlooked. You, however, have an opportunity to wrap your mind and actions around the following realities. In doing so, you can be in the top percentile of graduating engineers who will receive the best job offers by graduation, while being ready to launch a career of purpose, fulfillment, and financial success.

REALITY #1: YOU ARE IN CHARGE OF YOUR LIFE AND YOUR FUTURE

Observation

Most people don't have a high-flying ambition or a life plan. That's okay, especially if they are law-abiding citizens who are making a positive contribution to society. This chapter and this book are not urging you to be a workaholic. The hope is that you will thoughtfully choose the engineering professional goals that are right for you, blending your work, family, play, fitness, and other interests that will lead to a healthy, balanced life.

Aspiration

Webster's dictionary defines aspiration as "a strong desire to achieve something high or great." Hopefully, you will choose to live a healthy, balanced life, and you will also fulfill your professional potential. Professional success and a balanced lifestyle can lead to real happiness. This book is designed to give you the information to choose from many functional opportunities and operate at any professional level suited to you. Having a great *attitude* and the ability to take *action* toward your *goals* are three word-essentials of success at whatever you do.

Purpose

The happiest, most successful people are those whose lives are filled with purpose. Having purposeful objectives can fill each day with meaning and pleasure as you continuously progress toward "all that you can be." So says Dr. Martin Seligman, Director of the University of Pennsylvania Positive Psychology Center, former president of the American Psychological Association, and leading expert on fulfillment and happiness (www.authentichappiness.sas.upenn.edu).

Dr. Seligman's research shows that the happiest people are those who are living a meaningful life—a life of significant contribution. The happiest people make daily use of their best skills and passions, directed toward meaningful goals that benefit others. They are having fun by using their strengths and being absorbed in projects that they know are making a positive difference. And they get *paid* for having *meaningful fun* that uses their *natural talent*!

Regarding purpose, you, as an engineering student, are very fortunate; why? You probably chose to study engineering for good, logical reasons; that is, you were good at math and science and liked them more than other subjects. You selected engineering as the best profession to match your skills and interests. Great choice! Engineering is an extremely purposeful profession, taking scientific principles and building something useful. That useful result can be a structure, product, or process that solves problems and creates opportunities.

Your engineering contributions can be extremely valuable at many levels—your planet, your country, and you.

Your Planet

The Earth needs:

- ❖ Sustainable energy and environmental solutions
- ❖ Agricultural technology with food distribution systems
- ❖ Advanced bio-medical technology and life sciences
- ❖ New energy- and environmentally efficient transportation systems
- ❖ Ever-advancing computer and communication systems
- ❖ Advanced structures—buildings, bridges, highways, railways

This is just a start—you can add to the list.

Your Country

The United States needs:

- ❖ Leaders in the global imperatives described above.
- ❖ A strong technical industrial base—whether it's making products, processes, or information systems. An industrial base creates real

wealth, and it's the foundation of a country's standard of living. We look to engineers to create technology and products. Those products and technology provide jobs in the United States. The products are sold domestically and internationally to make our economy strong. The growing service sector of our economy is important for our many needs; however, the service sector does not make products or export them.

- A technically educated workforce—the world is becoming increasingly technical. That means there is great opportunity and need to teach and use engineering subjects. A career as an engineering professor can be meaningful and fulfilling.

You

Because you are in charge of your life and your future, you and your career are profoundly important. You can really make a difference. Consider this: many, perhaps most, people move through life without much in the way of goals or purpose. They're waiting for something to happen. Waiting to be told what to do. They're not the leaders or the happiest of people. In order to be the best you can be, you first need to know that you are in charge of your life.

Read on to find out how to accept the personal responsibility to create your future.

- Dr. Martin Seligman's research shows that the happiest people are those who are living a meaningful life—a life of significant contribution. Your meaningful contribution to society as an engineering professional can be huge, as well as giving you happiness.
- Dr. Abraham Maslow was the founder of Humanistic Psychology. Like Dr. Seligman, rather than study mental illness, Dr. Maslow's interest was in what constituted positive mental health. Dr. Maslow proposed a five-level hierarchy of needs in his 1943 paper, "A Theory of Human Motivation." At its foundation are basic survival needs—food, clothing, and shelter. At the top, and most relevant to you (we'll presume you have survival under control), is self actualization, that is, reaching your full potential: it's making the most of your abilities and doing the best you can. Self-actualization can be your throughout-life passion: learning, discovering, and growing to be your ongoing best. Reaching your potential can be your permanent positive motivational need and your source of energy to keep you successfully in charge of your life and your future.
- Dr. Mihaly Csikszentmihalyi, who Dr. Seligman has described as "the world's leading researcher on positive psychology," introduced the concept of "flow state" in his book, *Beyond Boredom and Anxiety* (Jossey-Bass 2000). As the book title indicates, there is a zone above boredom and below anxiety/distress called flow. In the flow state, you are immersed, totally absorbed in your work. You are energized, focused, engaged, so much so that you may be unaware of the passage of time. You're at your best! Do you get the feeling from this that you're having a good time? Yes, says

Dr. Csikszentmihalyi—you are in the flow state, happily absorbed by maximizing your personally combined levels of skill and desire for challenge. (Note to the engineer in flow state who also has a family: you may want to set an alarm for when it's time to go home so you don't miss dinner.)

- Big point, missed by many: you have control of your purpose and choice. Purpose and choice are yours, and the opportunities are there for you to create! You're on course already by choosing engineering as your well-suited, purposeful field of study. Continue to adhere to your purpose wisely and you're off to a great start, destined for happiness and success.
- Engineers build. You will be building your education and career, which will help people and the economy and create your own wealth in every aspect. These aspects include professional accomplishment, technical fulfillment, financial security, and the joy of living a significant life.

Summary of Reality #1: You Are in Charge of Your Life and Your Future

- You can take charge of your life and future. Unusually good or bad luck is not likely to enter the big picture. Chances are remote that you'll win the lottery or be struck by lightning. Your future can be in your hands, if you choose!
- You can, intelligently, work yourself up to ongoing self-actualization and spend your working days in an energized, desirable flow state.
- The reward for making the most of your potential can be living a fulfilled, meaningful life—especially if your work is also in balance with other important priorities such as family, play, and fitness.
- Ongoing happiness and sufficient wealth are the likely by-products of you "doing the right things" with your career and life. Our country really needs your engineering talent. You can, by continually building your skills and passions, evolve throughout life to be where you want—technically, professionally, and personally.

REALITY #2: SOFT SKILLS ARE NEEDED TO SUPPLEMENT YOUR TECHNICAL SKILLS

What are "soft skills"? They are the non-technical skills—that is, communication, leadership, teamwork, planning, time management, and other areas of competence that allow you to be successful in an academic, corporate, government, or any professional environment. Often, these skills are difficult or overlooked by very technically focused engineers. But, communicating, leading, and working in teams and with customers are crucially important to be a successful engineering professional.

Soft skills include emotional and perceptual intelligence, which originates from the right hemisphere of your brain. Logical, analytical intelligence is used in engineering problem solving and comes from the left cerebral hemisphere. Clearly, to be your career best and of maximum value to your employer, you need to have both left and right sides of your brain working effectively.

These "major research findings" are evidence of the importance of business success soft skills:

PREPAREDNESS FOR PRACTICE*

Respondents were asked to rate new engineers' preparedness for practice in eight areas and then indicate the value their organization places on preparation in that area. [Figure 2-1] shows the results for each area.

With the exception of "Math and Science," there appears to be a wide discrepancy between the value expectations of the employer and the extent to which their employees are seen to be well prepared. This would further appear to reflect on the mismatch between curricular emphasis and employer expectation. It must be recognized, of course, that math and science are, without argument, the key ingredients—at least in the lower division—of an undergraduate engineering education.

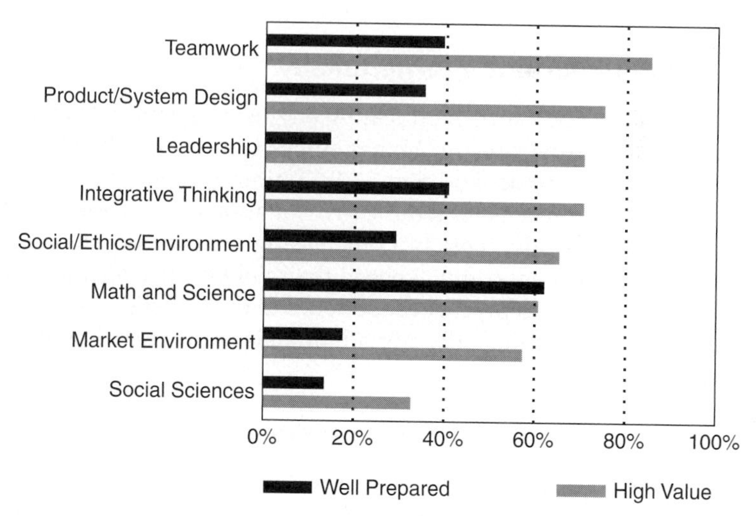

FIGURE 2–1 Value vs. Preparedness

These research findings are, more than ever, valid today. For proof of this, just Google-search "engineering soft skills" and look what comes up: hundreds of articles, speeches, and courses—all designed to emphasize the reality that it takes a person with a well-functioning right brain (perceptual IQ) as well as left brain (analytical IQ) to reach full potential and value to an

*From "First Professional Degree Survey Report," *Engineering Education Issues: Report on Surveys of Opinions by Engineering Deans and Employers of Engineering Graduates on the First Professional Degree*, NSPE Publication No. 3059 (November 1992): 5.

organization. Accordingly, this book contains important information for engineering students who expect to reach their full professional and life potential.

Employers Place a Premium on Soft Skills

Ted W. Hissey, Director Emeritus of the Institute of Electrical and Electronics Engineers (IEEE), has said, "today's companies place a premium on individuals who develop, practice, and continue to improve certain extra, or 'soft' skills . . . In the past, many engineers preferred taking individual responsibility for developing a product . . . today, corporations want individuals who exhibit strong teamwork, global perspective, and multiplexing capability." Mr. Hissey based these observations on "interviews with industry executives and managers, industry-savvy government leaders, and academic leaders from around the world . . . The consensus results indicate that engineers and scientists should understand the career-enhancing value of 'soft' skills to progress in today's global, open market economy."

Mr. Hissey concludes, "Highly successful professional engineers are not only technically astute, but also possess some of the extra, or 'soft' skills that many experts believe are becoming more critical for engineers and scientists today."*

Soft Skills Are Important Internationally

Recognition of the importance of educating engineering students in soft skills is truly global. G. Hilmer of the Management Center, Innsbruck, Austria, presented a paper on "Social and Soft Skills Training Concept in Engineering Education." The summary is "Scientific investigations have shown that . . . in today's industry and trade, there is an increasing demand for engineers who don't just have an excellent competence in their field of specialization but also a good understanding and practical experience in the so-called 'social and soft skills'. These subjects are usually not adequately addressed in engineering degree programs."†

Learning Soft Skills Outside the Classroom

There is so much required technical coursework in a four-year engineering curriculum that few elective courses are available to develop soft skills. Therefore, as suggested in Chapter 1, you are strongly encouraged to join engineering organizations or other activities, such as volunteer work, that will allow you to develop your leadership, teamwork, and communication skills. Co-op and internship assignments are also excellent means of applying your technical coursework while learning interpersonal and other business-related soft skills.

---

*Ted W. Hissey, "Enhanced Skills for Engineers," *Proceedings of the IEEE* 88:8, (August, 2000).

†G. Hilmer, "Social and Soft Skills Training Concept in Engineering Education." Paper presented at the International Conference on Engineering Education (ICEE), September 3, 2007, in Coimbra, Portugal.

It is especially important to seek an experience abroad, as described in Chapter 19, if you have an interest in working globally. These "co-curricular" (beyond the standard classroom) learning experiences will pay off by making you the totally qualified person that employers seek.

Summary of Reality #2: Soft Skills Are Needed to Supplement Your Technical Skills

- Soft skills are the non-technical skills. They include planning, time management, and people skills—communication, leadership, and teamwork.
- Research findings show that employers consider new engineering graduates very well prepared in math and science, but poorly prepared in soft skills—especially leadership, teamwork, and market environment.
- Soft skills are a necessary career enhancer for success in our global economy.
- You need soft skills as well as engineering technical skills to get the best job, to be promoted, and to advance to your full potential.
- Seek co-curricular opportunities to learn soft skills outside the classroom, through teamwork and leadership in engineering clubs, volunteer work, and career-related co-op/internship experience.

REALITY #3: YOU ARE A SALESPERSON

Most people do not have the job title of salesperson, but in fact, everyone's in sales whether they know it or not.

As an example, during my career course orientation I ask students, "How many of you want to get career-related experience, such as a co-op assignment or internship before you graduate?" All hands go up.

Then I ask, "How many of you expect ever to become involved in some form of sales?" No hands go up.

Then, "Since everyone wants an industrial co-op/internship experience and no one expects to be in sales," I ask, "how many of you have in mind an employer—other than your parent—who, out of the kindness of their heart, will charitably take on the expense of employing you just because you need job experience?" Pause—no hands go up.

After a brief discussion, the group realizes that if they are going to be employed, every one of them will be in sales. If you want to be employed, you'll be in sales, too. Specifically, in seeking employment:

- You are the salesperson.
- You are the product/service being sold.
- You should know that a business needs to show profit; employers won't incur expense without expecting value.
- Your job is to present persuasive evidence to the employer that you are a value-creating person and that hiring you will be a good business decision.
- You need to sell the employer that you will be an asset, not an expense.

You, as an engineering student, probably don't think of yourself as a potential salesperson; most engineering students don't. In fact, that may be one reason you chose to study engineering. Moreover, most engineers are focused technically and don't want to "blow their own horns," promoting themselves on resumes and in interviews.

Now the good news: the above observations are to your distinct advantage. Why? Because knowing you're in sales and using the points above put you at a distinct competitive advantage—most engineering students haven't given these ideas much, if any, consideration. Putting these concepts into practice will place you in the top 5–10% of job seekers. In the following chapters, we will show you how to market your credentials—how to sell yourself to the right employer, one who you decide is best able to use your skills, interests, and values. More good news: the process will not be painful. Using a logical process of self-evaluation, resume and interview preparation, and your market-niche research, you'll be surprised at how exciting your journey will be toward a "best job for you." You will be prepared to sell yourself, with evidence, in interviews. Closing the sale will be a thrill.

A good example of an engineering intern as a salesperson is Rick Licursi. You'll meet Rick Licursi in Chapter 13 when he teaches you value engineering. As an intern after his junior year, Rick used value engineering principles to sell his management on the idea that they needed to purchase a machine costing $110,000 to cut production waste. Would a company normally spend $110,000 on a suggestion from an intern? Perhaps not—but let's see what happened: Rick wrote a proposal and made a speech to his management to convince them of his idea. Rick's proposal paid off: the $110,000 investment paid for itself within six months. That means a mechanical engineering intern, following his junior year, made a recurring savings for his company amounting to approximately a quarter of a million dollars per year. Rick Licursi's job title had nothing to do with sales. But Rick had a cost cutting idea, and if it was not for his ability to sell his idea to management, it would've gone nowhere. So it will be for you throughout your career. Most engineers don't have the title of salesperson—but to be most successful, you need to be able to persuasively present your ideas in order to have them accomplished.

Summary of Reality #3: You Are a Salesperson

You will be in sales throughout life—to obtain employment, to have your technical ideas accepted, to be promoted. Prepare yourself.

- Most engineering students don't want to be a salesperson.
- You and every professional are in sales in order to get a job and be successful.
- You must thoroughly know yourself and your potential employer.
- You must connect your specific skills/experience with an employer's needs.
- You must prove, with evidence, that you will create value for the employer.
- Evidence sells! It's what you bring to the employer, not your need for a job.

Chapter Summary: Takeoff Tips

The three realities in this chapter are very important. Embrace them.

- ❖ You are in charge of your life and your future.
- ❖ Soft skills are needed to supplement your technical skills.
- ❖ You are a salesperson. You must present evidence that you are a value-creator.

Exercises

1. State the three realities and what they mean to you.
2. Describe why each one is important to you in your professional and personal life.

Looking Ahead

In Chapter 3, you will assess your salable experience, strengths, interests, and values. This combined self-assessment has major benefits to you. It will allow you to:

- ❖ Know, specifically, what you have to sell about yourself
- ❖ Direct your job search to the ideal employers for your skills and interests
- ❖ Have everything you need to produce a great resume and prepare for interviews
- ❖ Be confident in interviews, knowing you have evidence to support you
- ❖ Create a career and life plan that is a key to your success and happiness

CHAPTER 3

Self-Analysis: Your First Career Step

"If we did all the things we are capable of doing, we would literally astound ourselves."

THOMAS EDISON (1847–1931)

In Chapter 2, you were invited to take control of your life and create your purposeful future. You were shown that you need your soft skills as well as your technical skills and that you need to sell yourself in order to land jobs and promotions.

Typically, a job seeker first sits down and quickly produces a resume, with no particular objective in mind and most likely not geared to really sell their best experience, strengths, interests, and past results. You are better than the typical job seekers just described.

Aim for and land a position in the correct niche in the job marketplace—a niche that is just right for you. Sound logical? That niche will allow you to use your best skills, feed your greatest interest and passion, and allow you to demonstrate your personal values. The result will be the right choice of a job and a successful career. According to Dr. Martin Seligman, good choices and actions can lead you to genuine career happiness: a pleasant life, a fulfilled life, and the highest level—a meaningful life.

That's the general picture; now what's with self-analysis? Why and how are we going to do it?

WHY IS SELF-ANALYSIS IMPORTANT?

You'll get the ideal job only by intelligently transferring your best skills, greatest interests, and strongest personal qualities and values to the career marketplace. As a first step, inventorying your unique personal attributes is hugely important. This is true for your first job search and, periodically, throughout your career. Why? Inventorying, then selling your transferable skills on an ongoing basis, is what career success is all about. Doing this ensures that you're in the right place, doing the right thing, at the right time. Only you can steer your career course. Choose to thoughtfully do the self-assessment in this chapter and it will pay off.

To sell yourself to an employer, you need to first decide what experience, skills, interests, and values you have. Then decide which employers you are going to approach for a win/win relationship—where you and the employer both benefit. Most job candidates don't take the time to do this, giving you a decided competitive advantage. You will know yourself and what you have to sell to an employer of choice. You will be able to present persuasive evidence that you are competent, motivated, and possess the personal qualities that your prospective employer needs.

HOW TO DO SELF-ANALYSIS

The following exercises will allow you to logically take stock of yourself and build factual evidence that you will use in your resume and interviews. Your self-analysis information will be invaluable for targeting the right employers and presenting yourself to them in the best light.

Visit the Companion Website at www.pearsonhighered.com/millar to conveniently do the self-analysis exercises using your computer.

SELF-ANALYSIS EXERCISE

A. Academic Assessment: List on a separate page.

1. Which courses do you especially excel at? ____________________
2. Which courses do you like the best? ________________________
3. Which were your worst courses? (skill/interest) _______________
4. Are you competent in engineering theory? (specify) ___________
5. Are you competent in hands-on application? (give examples) ____
6. What is your engineering major/minor? ______________________
7. Do you want to work in your major or some other discipline? (specify) ______________________________________

B. Honors/Awards/Scholarships: List academic, leadership, or other significant awards.

1. ____
2. ____
3. ____

C. Hobbies: List hobbies or other spare time activities that help define your skills and interests.

1. ____
2. ____
3. ____

D. Specialized Skills

Computer

1. Applications in which you excel ______________________
2. Applications of most interest to you ______________________
3. List all of your computer skills on a separate page ______________
4. What is your level of interest in a job with extensive computer use? ______________________

Specialized Equipment/Processes: Specify for each, as appropriate.

1. Lab equipment ______________________
2. Machinery ______________________
3. Processes ______________________
4. Other ______________________

E. Experience Assessment: To complete the experience assessment, make a separate sheet for each significant employer or organization. Especially focus on your demonstrated skills, strengths, results (be specific, quantify if possible), likes, and dislikes. This exercise will give you valuable information regarding what relevant experience you will decide to offer to an employer.

Each sheet can summarize a position with a company, a student organization, or any activity that you feel might make you valuable to the marketplace. The experience does not have to be technical, especially if leadership, teamwork, business, or interpersonal skills were involved. You will want to especially discern what you were good at and what you liked doing. Use a felt-tip marker to highlight these positive items; they will be important to consider in forming your objective and resume. Look for a pattern of areas of competence and strength; these will be keys in formulating your professional objective.

Experience: Make a listing of each employer or organization in which you have held a job, an office, or a leadership role, summarizing the following:

1. Your position ______________________
2. Your duties ______________________
3. Your specific skills ______________________
4. Your strengths ______________________
5. Your results, including recognition ______________________
6. What you liked ______________________
7. What you disliked ______________________

F. Personal Qualities

1. Are you good at working on teams? (examples) ______________
2. Do you like working on teams? (explain) ______________

3. Do you prefer working independently? (explain) ______________
4. Are you at your best working alone? (examples) ______________
5. Are you a good leader? (examples) ______________
6. Do you like being a leader? (explain) ______________
7. Do you have personal requirements regarding location, income, or other needs? ______________
8. What work should you avoid? (lack of skill or interest) ______________
9. What especially motivates you? (specify) ______________
10. What are your professional skills/strengths? (that relate to a job objective) ______________
    - ______________
    - ______________
    - ______________
11. What are your professional interests? (specify with examples)
    - ______________
    - ______________
    - ______________

G. **Overall Accomplishments:** List your most important (to you) accomplishments so far in life that best exemplify your strengths, interests, and experience (specify and explain on a separate page)

   1. ______________
   2. ______________
   3. ______________

CAREER ANCHORS

The previous self-analysis exercises will be crucially important to planning your future success and happiness. Now you are more aware of the talents, interests, and values that will determine an ideal career placement for you.

Now, before we move on, let's examine the useful application of your self-assessment alongside the "Career Anchors," developed by Dr. Edgar H. Schein, Professor of Management Emeritus at MIT and pioneer in the field of career development.

Your next step is to identify your responses and patterns in the analysis that you just completed in order to help you identify one or more career anchors, or clusters of your self-perceptions. Based on Schein's descriptions of career anchors, decide where you belong at this point in time based on your skills, interests, and personal qualities/values/goals. You can fit into more than one category, and your desired category may change over time.

Characteristics of Career Anchors: Where Do You Belong?

Technical/Functional

- ❖ Need to emphasize technical specialist skills in engineering.
- ❖ Not primarily interested in managing people.
- ❖ Enjoy being an expert in a specific area.

Managerial

- ❖ Need to plan, organize, lead, and control the work of others.
- ❖ Interested in using interpersonal skills with individuals and groups.
- ❖ Enjoy power and responsibility.

Entrepreneurial

- ❖ Enjoy risk taking and overcoming obstacles.
- ❖ Enjoy being personally responsible for results.
- ❖ Need to create, innovate, build, or develop new ways of doing things.

Security/Stability

- ❖ Want a secure, fairly predictable environment.
- ❖ Enjoy knowing that job and financial security are stable.
- ❖ Low interest in changing lifestyles, moving, or starting something different.

Autonomy/Independence

- ❖ Desire to be free of organizational rules and constraints.
- ❖ Enjoy making own decisions about what, where, when, and how to work.
- ❖ Remaining independent is more important than a promotion.

Service/Dedication

- ❖ Need to contribute to making the world a better place.
- ❖ Enjoy helping others more than seeking positions of power and wealth.
- ❖ Enjoy providing value to others' lives.

Source: Edgar H. Schein, *Career Anchors: Self-Assessment, Third Edition*. Hoboken, NJ: John Wiley & Sons, Inc., 2006. Reproduced with permission of John Wiley & Sons, Inc.

TARGETING YOUR IDEAL CAREER OR STARTING JOB

Look over your self-analysis. Examine your special skills, interests, personal qualities/values, and overall accomplishments. See how they may fit into one or more career anchors for you. Generally, the things you do best are the things you're most interested in, and vice versa. Most people don't excel at things they don't like—you're not a par golfer if you hate golf.

Let's proceed to target an ideal job for you. The right career choice can, as has been stated, produce a career of fulfillment (Maslow), flow state absorption (Csikszentmihalyi), and happiness (Seligman). Realistically, life is not perfect, and we don't want to elevate expectations beyond reason. That said, the assessment you have completed and the written goals you are about to set will likely pay off with much more success and career satisfaction than ignoring a career plan and waiting to get lucky in the hope of a great career dropping into your lap. There is much evidence (e.g., Maxwell Maltz MD, *Psycho-Cybernetics;* Dr. Michael LeBoeuf, *Working Smart*) that people who have assessed themselves and set written goals achieve far more in their careers than those who have not.

It is most important during your first years out of school to obtain foundational experience you can build on in the direction of your career choice. In the case of an engineering career, you will likely want to apply your engineering coursework in a technical job that matches your strengths and interests. You are building your track record for future success. Chapter 4 describes this in greater detail.

Now, list, in order of priority, what you want in your ideal job. This will lead to defining your resume objective and the employers you wish to target. Include "must haves" first, then "want to haves."

- ❖ ______________________________
- ❖ ______________________________
- ❖ ______________________________

YOUR JOB/CAREER/LIFE OBJECTIVE

At this point, it is logical to summarize your self-analysis results in a goal statement of what you want in the immediate future, as well as your long-term career. Write down what you want out of your career and life. Write down the realistic details that, if possible, will give your life the most satisfaction and define your success. You are creating your future by writing it down. A written plan will become an action plan for your success!

Read the following and state your desired function, type of employer, and type of market served. Include your short- and long-range goals. "My career objective is to apply my best skills/strengths/interests/values toward something I really believe in doing. Specifically, that something is" ________

____

____

____

Your self-analysis and goal statement are accomplished. Now, you are far more prepared to write your resume and to land the job you want than most of your competitors. Your careful self-assessment, made using the suggestions in this chapter, will enable you to effectively "sell yourself" to an employer who needs your skills to solve a problem or pursue an opportunity. This detailed career plan is enormously important, even though your immediate job objective statement will be condensed to 1–2 lines on your resume. Self-analysis, your most important work, is already done!

Don't worry about being too fixed into a career plan. As noted, this initial plan will evolve—technically, professionally, and personally throughout your life. It's great to have a plan, and the substantial effort invested in this initial plan will establish a firm foundation for your continued career growth. Planners win. Winners plan!

Important note: Your Career Center can offer additional tools and counseling for your self-assessment. The Myers-Briggs Type Indicator, the Self Directed Search, the Strong Interest Inventory, and DISCOVER are all examples. They encourage introspection and self-examination of your personality, interests, and skills relative to career plans. Your choice of using these Career Center services can contribute to your best personal and professional decisions.

Chapter Summary: Takeoff Tips

Self-analysis of your unique personal attributes is essential for your resume:

- ❖ Academic assessment
- ❖ Honors/awards/scholarships
- ❖ Specialized skills
- ❖ Experience assessment
- ❖ Personal qualities
- ❖ Overall accomplishments
- ❖ Career anchors
- ❖ Targeting your ideal careers/starting job
- ❖ Your job/career/life objective

Exercises: Self-Analysis

1. Do your self-analysis on pages 23–25.
2. List, in descending order of priority, what you want in your ideal job.
3. State your job/career/life objective. "My career objective is to apply my best strengths/interests/values toward something I really believe in doing. Specifically, that something is ________."

Looking Ahead

Next, in Chapter 4, we are going to do market research of engineering opportunities to decide where and in what function you want to work. Where is your market niche? Who needs your talent/passion?

CHAPTER 4

Market Research: Engineering Opportunities

Your effort during Chapter 3 in defining your skills, interests, and values into a tentative career plan is really going to benefit you. In this chapter, we'll find out how when you link your personal profile with the wide variety of engineering opportunities. Let's outline a sequence of information that can be helpful to you in getting your arms around some major engineering career options.

- ❖ Your foundation experience
- ❖ Some major engineering functions/career paths
- ❖ Hot fields—current opportunities and trends
- ❖ Types of employers where you want to work

We are proceeding to target one or more areas of opportunity that best match your talent and work you can love.

YOUR FOUNDATION EXPERIENCE

At the end of the last chapter we made two general suggestions:

1. It is most important during your first years out of school to get foundational experience you can build on toward your longer term objectives. Get the experience that will confirm your career direction. Explore your options. Head in the direction you initially selected, or make an intelligent change. Build your skills. Recognize what you do well and really love. Start establishing a great track record of accomplishment.
2. You should apply your engineering coursework in a technical job that matches your strengths and interests. The experience of applying engineering theory on the job is of major importance. This is particularly relevant to consider when you may face two attractive job offers: one more technical and one less so. If you are at the coin-tossing stage of which job to choose, go for the more technical one. Why? You will start

to forget engineering coursework if you don't use it soon. The bottom line is that you can easily go from more technical to less technical assignments, but the reverse is difficult—unless you want to go back for more technical education. Beginning in research, development, or design and then going to sales or management is technically easy. Going from sales to research is unlikely.

ENGINEERING FUNCTIONAL SPECIALTIES

It is important to define major functional engineering specialties. We are doing this now so that you can understand them and visualize yourself working in one or more of these key functions as co-op/intern students, new graduates, or in future years. How do your skills, interests, and values line up with these functions? Do you have evidence from your coursework, experience, personality, or aspirations that make one or more of these a good fit for you? The functions from the most highly technical/theoretical to the most general will be described.

RESEARCH—THE LONG-RANGE VIEW

Engineers working in research are closer to pure theoretical science than other functions. They are looking anywhere from one to ten years down the road to discover a "next generation" product or technology. If you are a researcher, you will dream of future applications of existing technology. You may also create a whole new technology or markets that are non-existent today. Research engineers work in universities, industries, and government.

The Work

- *Conceptualizing.* The research engineer's creative mental process of originating a new technical idea. "Here's something worth exploring!"
- *Analysis.* Researchers use engineering theory to mathematically analyze the functional possibilities of their idea before going further.
- *Laboratory Experimentation.* Potentially valuable breakthroughs that have been derived mathematically need to be tested. With assistance from lab technicians, research engineers build and use equipment that allows observation of theoretical predictions. Experimentation results will prove, disprove, or allow improvement of a dream and analysis.

Qualifications

Research engineers will generally have advanced degrees, commonly a PhD, and typically work in university and large industry settings. They have high grade averages, with superior skill in engineering analysis and laboratory experimentation. Prime qualifications include creative imagination, curiosity, and technical energy to pursue a previously unknown concept into reality.

A researcher should have the patience and tenacity to pursue long-range goals. Research is not for those who need to see immediate results. Legend has it that Thomas Edison was once asked how it felt to have failed at over 900 attempts to discover a filament for his dream of an incandescent light bulb. His response: "Failure? I have *discovered* over 900 filaments that don't work!" Wilson Greatbatch, inventor of the implantable pacemaker and holder of over 325 patents, says, "Nine out of ten of my ideas don't work, but the tenth idea pays for the other nine."

DEVELOPMENT—SCALING UP FROM RESEARCH TO DESIGN

Research and development are often grouped together, commonly known as R&D. The functions are a bit different, so they are described separately. After a researcher has discovered a potential product or process, the next step is development. Development engineers take a concept from research and scale it up into a marketable reality. The technology is still in an early stage, but what was a research possibility years ago is now being brought to life with a targeted customer in mind. Therefore, development is the step between research and design.

The Work

- ***Analysis.*** A concept proven by research to be workable needs to undergo mathematical and customer need analysis to prove that it is marketable on a technical and cost basis.
- ***Laboratory Testing.*** Likely a prototype or pilot plant, it is the first experimental, small-scale version of what will become the actual product or process that will be later sold to a customer. By the time it leaves development, the technology will be "scaled up" to a workable configuration that can be designed, manufactured, and sold.

Qualifications

As indicated, research and development (R&D) in industry, government, and consulting are often linked together in one department. Therefore, the qualifications are often similar: advanced degrees and high grade averages. The development engineer, however, is more likely to have an MS degree as compared to the PhD research engineer.

Development differs from research in that there is an immediate, focused market application for what was a long-range research concept. Therefore, there is more specific attention to the potential customer requirements: use, cost, and quality. Development engineers work with, or may be, researchers while also communicating with engineers who will be designing the product/process. Their skills are highly technical. Generally, however, development engineering skills are less abstract than research skills and more practical, with more communication interface with other groups. Development engineers have to get a project moving into the design phase within a time frame, as opposed to the researcher who is creating technology that previously did not exist.

DESIGN—ENGINEERING THE END-RESULT

With developed technology and a customer application in mind, design engineers will individually, or more often on a team, produce a specified design of a product or process that can be manufactured. The design team, especially in large organizations, will often include members of different engineering disciplines, depending on the customer and technology mix required.

The Work

Design is increasingly done as a multi-discipline team function. Members of the team contribute calculations and specifications relevant to their function or discipline. As an example, for a chemical process plant, a chemical engineer would do the overall process and control system design; a civil engineer would do the foundation design; a mechanical engineer would do the distillation column and piping design, as well as the mechanical equipment specification; and an electrical engineer would do the power equipment specification and distribution, as well as the control systems. A project manager or design team leader would ensure that everyone on the team is coming together technically, on schedule, and within budget to ensure project success.

Individually, you, as a design engineer, might direct a designer or drafter to help produce your design. Now, with CAD and other solid modeling software available, you might do the designing yourself. In smaller organizations or projects, you might be the only one doing design, despite the larger team trend mentioned above. Whether you are the sole designer or are contributing on a larger team, you will be interfacing with a number of other people in functions such as R&D, purchasing, and manufacturing. You may travel to a customer's location, including a plant start-up, to ensure your design is operating as intended.

Qualifications

Design engineers will have a BS degree, and sometimes an MS degree, with good grades and the ability to apply engineering coursework to produce a design that meets customer requirements. They will have knowledge of CAD, Pro/ENGINEER, or other software to allow them to produce drawings and good design. As mentioned above, increasing emphasis is now being placed on designers to be effective team members with good communication and interpersonal skills, who are also technically sound.

MANUFACTURING/PRODUCTION—SUPERVISING PRODUCTION OF THE PRODUCT

Manufacturing or production engineers supervise plant workers in making the finished product meet customer specifications. Manufacturing has become increasingly automated and computerized. Nonetheless, a manufacturing or production engineer is likely to be a team leader or supervisor of production workers.

The Work

A manufacturing or production engineer is responsible for managing a diverse workforce in his/her area of responsibility, including the functions of planning, organizing, leading, and controlling. Schedule, budget, and quality are of great importance. Accordingly, Six Sigma/Total Quality Management and Lean (to be described in later chapters) are in widespread use. As a manufacturing supervisor, you will have the opportunity to learn and demonstrate management skill with broad focus on your team and a production process. You may also, with globalization, be traveling to other countries to ensure the product is being made to specification.

Qualifications

Typical entry degree levels are BS and, sometimes, MEng, MS, or MBA. Leadership capability is a prime qualification for manufacturing or production engineering. You should enjoy the challenge of leading a diverse team of production workers and seeing a quality product being built within budget and on schedule. You need to be technically competent, but with emphasis on hands-on practical problem solving skill, rather than being an ace in engineering theory.

TECHNICAL SALES—SOLVING CUSTOMERS' NEEDS WITH YOUR PRODUCTS

While engineers generally may have disinclination toward sales, it can be an excellent functional area of career opportunity, particularly if you have already had some previous opportunity to apply your engineering coursework to develop or design a product. The key opportunities in sales are breadth of exposure to your company's product line and the range of potential customers and their technology needs. Sales engineers are generally well paid. The reason: they can make a very positive, measurable impact on the bottom line of their company's profits. The company doesn't make any money until its products get sold. Is sales a good fit for you?

The Work

Sales engineers identify potential customers, phone for appointments, and then travel alone or with another tech-support person to those organizations in their territory that could profit from using their company's product. The competent sales engineer will grasp the potential customer's technology, including their mission of technical/business improvement, and relate it to what their company can provide. The sales engineer's goal: making a persuasive presentation that will convince the prospective customer that it is in their best interest to buy the product. To accomplish this, a sales engineer needs to be technically competent in both their own and the customer's technologies and to show how the customer can profit by purchasing the product. The sales engineer will then negotiate a contract and arrange for administrative and technical support for the deal that has been made. Sales offers a lot of exposure to the employer's business and that of the customers they serve. Sales engineers may attend trade shows, technical conferences, anything that will put them in touch with the state of their technology and potential customers. This broad awareness along with interpersonal and business competencies can lead to management opportunities.

Qualifications

Presenting and selling their company's technical product to a customer requires two major strengths:

1. Thorough knowledge of the products and related technology applications.
2. Ability to quickly learn the customer's technical need and to connect that need to their product as a solution for the customer. Closing a deal is all about the company's profit and therefore a commission; that is why successful sales engineers are well paid.

Sales engineers typically have a BS degree and sometimes an MBA. They often work independently and out of their homes. Therefore, they need to be organized and self-motivated to get up in the morning and make every hour count toward sales goals. Functionally, sales engineering work includes searching for new clients and customer applications. Successful sales people are natural competitors who thrive on the challenge of making win/win deals. Performance is especially measurable, and sales engineers look forward to the immediate, quantifiable results that allow them and their company to "keep score." They are generally high-energy, confident people who welcome this environment and the opportunity to close as many winning deals as possible.

SECTION CONCLUSION

We have just outlined major engineering functions where you might work. Now think about your self-analysis and goals from Chapter 3. Are you starting to see some definition of what function you might want to get started in and where you might head for the longer term?

Again, during school, it is good to get some immediate experience doing actual engineering work, applying your coursework on projects that are as technical as your skills and interests allow. It will be helpful to write down a summary of thoughts at this point. What's right for you?

EXERCISE: Select one (or more) of the above engineering functional specialties that best suits your specific skills and interests. Describe:

Specifically, how do your skills match the function(s)? Why are you qualified?

____

____

____

Specifically, what motivates you toward this choice? Why would you be happy here?

____

____

____

HOT FIELDS AND CURRENT OPPORTUNITIES AND TRENDS

Let's give you more information about currently "hot fields" and trends. Afterward, we will describe different kinds of organizations and work that you can do.

By all means, don't jump into a hot field just because it happens to be hot now. You may or may not be qualified, and its current popularity may mean a lot of job applicant competition.

Throughout your career it's best for you to keep your eyes and ears open for trends and opportunities. Read newspapers and journals. Have conversations, but especially listen. Observe problems. Consider creative solutions.

Here are some thought starters describing some technical areas and trends that are currently drawing attention as special areas of opportunity. These are just a few examples, and areas of hot opportunity will continually emerge and change.

Energy and Environment

Energy and environmental solutions will represent great opportunities for engineers who can help with intense needs for new, efficient, and sustainable approaches to the challenges we face in the 21st Century.

More companies, as well as the government, are moving into these fields, commonly referred to as "green engineering," and will be doing so in the foreseeable future. There is a wide-open market for new ideas and economically feasible new technologies in these areas. Some engineers will become extremely successful by commercializing practical, new solutions to energy and environmental dilemmas. Why not you?

Biotechnology

Opportunities are rapidly increasing as medical science and engineering merge to capitalize on scientific breakthroughs and an aging population. As a trend, engineering application is broadening beyond the traditional boundaries of a specific major. In this biotechnical technology area, successful engineers will add knowledge of physiology to their engineering base of skills. Broader knowledge and increased contact with people, reading journals, and attending conferences that are at the cutting edge will increase your probability of making breakthrough contributions.

In order to be at the cutting edge of knowledge, you may decide to obtain one or more advanced degrees in your areas of biotechnical interest. Since many biotechnical companies are small, networking is often necessary to discover job opportunities; they may not appear in university-sponsored job postings.

Global Opportunities

The twenty-first Century has been marked by a huge increase in globalization of business and technology. Read *The World Is Flat* by Thomas Friedman for in-depth perspective of this area; the more you are aware of the internationalization of business, the better off you will be. Be willing to travel and relocate anywhere and you'll be in demand. Learn Chinese or another language suited to your targeted industry. Become international and there will be a career payoff for those who are willing to go the extra distance. Chapter 19 provides detail for you on becoming a global citizen.

Niche Markets

Another new phenomenon is niche markets and is well described in *The Long Tail* by Chris Anderson, a physicist and editor in chief of *Wired* magazine. He writes in detail about how numerous instant niche markets are surprisingly and successfully taking shape right now. Look at Google, Amazon, eBay, and Netflix. Technology is making mass markets smaller and niche markets expand. This very recent trend is all about the Internet being a market in search of products and the increase in consumption of niche products. Niche marketing is about the ability to sell to discrete, smaller markets. This is a major cultural and business environmental shift. Moreover, only a small percentage of the population really knows about it! Maybe you already do, and you might decide to take advantage of it. Read books like *The Long Tail* and *The World Is Flat* and you will be much better equipped to be at the cutting edge of what's happening, and then making something happen.

Things You Love

Do you love electronics? Motorized vehicles? Video games? A particular sport? Many do, but if you love anything with a passion and that activity lends itself to technical innovation, you might self-define it as a hot field for a niche market. Be aware that your niche may be crowded with others who have similar passion, especially if recreational by nature, but if you have a breakthrough, marketable idea, you may have a lot of successful fun and profit ahead of you. Development of video games is especially hot now, but you may have a better competitive chance if you are in love—or can become in love—with a niche that is less "sexy," more unknown. For example, you could start a hot market concept for an aging population. Always keep abreast of the news, trends, and things you can observe that might allow you to be among the first to spot a market niche need that is in line with things you love technically that aligns with your skills and values. Something that can become your passion! Bill Gates, Henry Ford, and the Wright brothers did—you know the rest of their stories. In each case—Gates, Ford, and Wright—there was no previous market or defined need—who knew? Now we regard their technologies as indispensable! What would we do without desktop computers, automobiles, and airplanes? Later, we will devote a whole chapter to becoming an entrepreneur and starting a business.

EXERCISE: Is there a hot field that you are passionate about and qualified for?

Name the field. (or your creative idea) ______________________________

What can you contribute? (your skill) ______________________________

____

Why are you interested in this? ___________________________________

____

Define the market. ____

____

What makes you think it will sell? _________________________________

____

Define the competition and your unique competitive advantage. (idea, cost, and so on) ____

____

TYPES OF EMPLOYERS—WHERE DO YOU WANT TO WORK?

So far in this chapter, we have discussed gaining foundational engineering experience before or after graduation. We have defined major engineering functions that might suit you, ranging from high-tech R&D to broader manufacturing and sales. We have also just described a sampling of hot fields that may interest you. Now, let's turn to types of employers where you might best be suited to release your talent and energy.

Large Corporations

Major corporations will be visiting your campus each fall in search of engineering co-op, internship students, or new graduates who are competent, motivated, and have the personal skills to fit their team. If you're interested, sign up early for an interview. What are the benefits of working for a large company?

- ***Lots of Options.*** Corporations have many professional areas in which you can contribute. You may have the opportunity to sample various functions and then choose a path or specialty that's best suited to your skills and interests. If you are successful, there are more levels to which you can be promoted—on either the management or technical ladder.
- ***Training Programs.*** Large corporations are more likely to have technical or management training and career development programs than smaller organizations. Through such programs you will receive exposure geared to optimize your career.
- ***Stability.*** These dynamic days, there is no such thing as guaranteed long-term job security. Nonetheless, working for a large corporation gives you a better chance that the organization will be there for the foreseeable future, as opposed to a small, start-up company.

Small Companies

Some small companies may visit your campus, but they are not as likely to have a recruiting staff to do campus recruiting on the scale of the biggies. There are advantages of smaller companies, however, that justify consideration.

- ***Breadth of Exposure.*** Drawing an analogy from the Navy, a junior officer's responsibilities on an aircraft carrier (large company), while important, would be more specialized. A junior officer's responsibilities aboard a destroyer (small company), would be considerably broader, with less formal training. A small company can offer you that breadth, along with a lot of flexibility to innovate. The president will probably know your name and what you are doing. You will know the people and their functions throughout the company. Small companies provide great opportunity to view the general picture of how an organization works.
- ***Growth.*** Most small companies are not going to take off as did Microsoft, Google, and Cisco. However, all large companies started out as small companies and if you pick a winner that is in line with your skills and passion, you might be able to ride that small company, through your technical and/or managerial contributions, up to the major league.

Government

As an engineering student, you might not immediately think of government as a prime field of opportunity. Then, looking more closely, you will see that there are many areas of technical need, challenge, and opportunity at federal, state, and local levels. They are looking for good engineers from all disciplines for important work.

- ❖ ***Federal.*** When we start listing federal organizations, you may agree that many of these deserve career consideration. All have important missions and significant technical and management opportunities. To name a few: Department of Energy, Environmental Protection Agency, Department of Homeland Security, NASA, Army Corps of Engineers, and Naval Civil Engineering Corps. Federal organizations offer many of the same benefits and opportunities as large corporations. You may have a passionate interest in energy, environment, aerospace, and so on that would make one or more of these organizations worth a good look. These organizations are often at the cutting edge of their technology.
- ❖ ***State and Local.*** The state and local government organizations also have opportunities for engineers majoring in most disciplines. Civil engineering is particularly relevant to state departments of transportation and public works commissions. If you have a preference to live in a particular state or locality, it will pay you to look over your opportunities to make an engineering contribution in that town or state. Engineering careers at state and local levels are generally less at the forefront of technology when compared to the federal level, but can nonetheless be very challenging and satisfying. They can be broader in scope, as in small company environments. The lower the level of government, the more you are directly in touch with the citizens. Your technical work may be connected with local public opinion considerations, such as proposed building or highway construction.

Education

As an engineering student, you are helping satisfy a much publicized need in the United States. The education profession needs engineers and scientists! Good professors will always be needed to teach engineering. Better math and science preparation is needed in elementary, middle, and high schools. There is opportunity as a technical educator at all levels.

- ❖ ***University Professor.*** Chapter 18 is devoted to academic careers and graduate school. However, let's say this now: if you achieve your PhD in engineering and love the academic environment of teaching, research, and publishing, being an engineering professor could be for you! By the time you earn your doctorate, you may have done sufficient teaching and research to decide whether the academic environment is the best fit for you. One thing is certain: a person well suited to an academic career will find it very satisfying.
- ❖ ***Middle or High School Teacher.*** If you love to teach and have the interest and ability to inspire younger people, there is a great demand for middle and high school teachers who really know math and science. That is one of the areas of greatest need in the United States. Average

pre-college math and science test scores in the United States have fallen behind many other countries. If you want to try out this career, add an education minor to your engineering degree; you will do practice teaching at your preferred age level.

Consulting

Consulting firms are hired by organizations to provide specialized expertise that they do not have on their regular payroll. Therefore, to be an engineering consultant, you need specialized experience that someone is willing to pay for, usually on a short-term basis. Where do you get that experience? You get it by first being employed in a function that will allow you to apply engineering fundamentals toward profitable project results. Establish yourself as a professional expert. Engineering consultants often have advanced technical degrees or an MBA. That said, many engineering students are hired at the BS level by consulting firms, often with the expectation that BS level hires will continue their education toward graduate degrees.

Other Opportunities

Once more, in most cases, new graduates are best advised to put their technical education to work on assignments that will allow them to apply engineering theory on a technical project. That allows them to establish their foundations as real, practicing engineers, just as a person earning their MD would logically go right into practicing medicine before doing something else. In summary, your engineering education is extremely marketable throughout the spectrum of technical as well as non-technical enterprise. Why? Did you notice that you have been studying harder on more difficult subjects than many of your fellow students? You have been developing an ability not only to understand technology, but also to analyze problems and derive logical solutions. Engineering's analytical skill is huge and sought by many industries, whatever their product or service.

- ❖ ***Financial Services.*** Engineers can be stock brokers, investment bankers, or financial planning consultants—in fact, you can find engineers in the whole scope of the financial industry. Reasons for this: they have strong analytical and problem solving skills, better than most others! Engineers understand technology; that is a big advantage in assessing investment in a technical enterprise.
- ❖ ***Non-Technical Consumer Business.*** You as an engineer are sought in many non-technical organizations for the same reason as just described in financial services. Every business needs professionals with strong analytical and problem solving skills. Some examples of industry segments seeking engineers are pharmaceutical, consumer goods, hospitality, and non-profit. Whether a company is making soap, candy, or prescription drugs, engineers are in demand for their analytical and technical skills that are necessary to produce the product. Let's not forget

that soft skills—teamwork, leadership, communication, and similar skills—are also needed in every category of enterprise: private and public, large and small, and technical and non-technical.

- ❖ *Law.* Patent law, in particular, is a field in which engineering knowledge is put to use in examining and processing technical invention applications. You would commonly enter this field after working as a functioning engineer. R&D experience is especially relevant, since you would be grounded in a creative, invention-producing environment.

EXERCISE: Which one or more of these organizations are right for you?

Your skills? ____

____

Your interests? ____

____

Your values? ____

____

Your career objectives? ____

____

CONCLUSION

Engineering is a superb profession with many functions and organizations from which to choose. Engineering is also a great place to be *from*. Your technical and analytical skills, combined with your soft skills of communication, teamwork, leadership, and other *Ready for Takeoff!* competencies will be in great demand for a wide variety of employment opportunities.

Once again, you are in control of your future. Be proud, confident, and motivated to prepare for it. Right now you have your self-analysis from Chapter 3 as well as engineering career option information from this chapter.

Chapter Summary: Takeoff Tips

Target areas of opportunity that best match your talent and passion.

- ❖ Study your self-analysis.
- ❖ Consider establishing foundational experience.
- ❖ Examine engineering functional specialties, including the work and qualifications.
- ❖ Assess hot fields—current opportunities and trends.
- ❖ Evaluate types of employers. Where do you want to work?

Exercises

1. Which engineering functional specialty is best for you? Why?
2. Which area of job opportunity interests you most? Why?
3. What will you do to use this information for your future?

Looking Ahead

In Chapter 5, we will put this all together into an excellent resume and cover letter. That combination will allow you to effectively present yourself to the market niche that suits you best. You are on your way to your choice of employer that will best satisfy your needs while you satisfy theirs. Yes, the name of the game is win/win! Both you and your employer are happy and profitable.

CHAPTER 5

Your Resume and Cover Letter

"The greatest danger for most of us is not aiming too high and we miss it. Rather it's that we aim too low and we reach it."

MICHELANGELO (1475–1560)

Writing your resume—seize the prize we've been leading up to! Now your diligence in studying material in Chapters 3 and 4 will pay off as, together, we prepare a winning resume for the career-launch position you deserve. Would you like to aim high to reach your professional potential? This chapter tells you how to do it. By chapter's end, I'll also tell you how to write an effective cover letter to accompany your resume.

Your resume is the most important personal, concise, results-oriented, sales document you will ever produce. You're now at a key point in your life-success journey. Therefore, it is important to review the last two chapters. These were very important assessment process steps that have prepared you to write a much better resume than most other job seekers.

In Chapter 3 you completed a thorough self-analysis. You determined:

- ❖ A professional understanding of yourself, your goals, and career objectives.
- ❖ The ideal combination of your skills, interests, experience, values, and goals.
- ❖ The personalized package of information required for your winning resume.
- ❖ Information you need to aim for the right market niche.
- ❖ Evidence of your accomplishments needed to convince the employer to hire you.

In Chapter 4 you did market research for engineering opportunities that could be best for you based on your self-analysis created in Chapter 3. This included:

- ❖ Your foundation experience
- ❖ A description of major engineering career functions—from research to sales
- ❖ Some hot fields—current opportunities and trends
- ❖ Types of employers—where might you want to work

Important to note: Your diligence in studying yourself and the marketplace in Chapters 3 and 4 are extremely important—and often neglected—process steps in a successful career launch. Most people, including engineering students, don't do this. That means most of your job competitors can't readily and knowledgeably describe who they are or where they're going. On the other hand, your new professional self-awareness and awareness of the marketplace will be to your big advantage. You will use this knowledge as you write a winning resume that will summarize your qualifications and attract employer interest in your background for the right job.

Also, you should be aware that some of the latest terminology being used in job search and career development is "promoting your brand" and "marketing your credentials." That speaks to the reality that you have, and will throughout your career development continue to possess, a unique "brand." That brand is your recognized combination of skills, passions, experience, and goals that distinguish you as an expert value creator. Your brand identifies you and makes you special and uniquely attractive to targeted employers. For more on this promotional concept, read *Building Your Personal Brand: Tactics for Successful Career Branding* by Randall S. Hansen PhD or *Be Your Own Brand: A Breakthrough Formula for Standing Out from the Crowd* by David McNally. Meanwhile, promoting your brand and marketing your credentials is what this chapter is all about.

Now let's proceed to the essentials of writing your winning resume. First, ask yourself what your objective is in writing a resume. Clearly, the first thing to come to mind is to get a job. Well, that's true, but getting a job is a project, and we will continue to take the project in logical process steps. Your resume will be excellent, but an employer is not likely to call you after reading your resume and offer you a job without seeing you first.

Let's always remember—the main purpose of your resume is to get an interview! That brings us back to you as a salesperson. Your resume is your personal sales document conveying your personal brand to the job market. You're selling yourself on paper in order to get that interview. How do you best do that?

Put yourself in the shoes of an employer: they are very busy and also need help from someone with specific skills that will create value for the company. They have a lot of resumes to go through, looking for the right person. How much time will an employer spend looking at each resume? Only 20–30 seconds. What are they looking for? What will catch the eye and positively separate yours from all the other resumes? Big question: what will you put on one page that will arouse the employer's interest to invite you in for an interview so you can go on to get the job offer? Your winning resume will:

- Tell potential employers concisely who you are and what you have to offer.
- Be visually appealing, with sufficient white space, effective font (Times New Roman or Arial), and type size (11 or 12).
- Be well organized and scannable, aided by bullets and bold headings, preferably one page—two at the most. Able to be read in 20–30 seconds.
- Be results-oriented, with evidence of your special abilities and accomplishments. Use past tense of verbs in describing experience. Quantify results when possible, using dollar amounts or percentages of cost savings, productivity increases, and so on.
- Make every word count. Use short phrases rather than complete sentences.
- Have perfect spelling and grammar. Get it proofread by a qualified person.
- Emphasize your strengths related to your targeted market niche.
- Motivate the employer to interview you for more information. Arouse their curiosity and interest, but save detailed descriptions for your personal meeting.
- Be the outline you will use during the interview for you to expand results in more detail. You will have a short, positive, results-oriented story prepared to tell for each line in your resume.

Invest enough time to produce a winning resume.

General advice from career center directors: Students have been going to school for years and achieving valuable experience; if they try to summarize this onto a one-page resume within an hour, they are not spending enough time creating it. That's great advice. It doesn't mean that you have to labor over your resume for weeks, but the business of condensing your qualifications onto one page is extremely important. You and your future career deserve enough focused time to produce the best document you can, so that you can be interviewed and land the right job for your skills and interests. The good news is that doing the analyses in the previous chapters gives you the information you need. You just need to summarize it effectively onto one page. A counselor in your career center can help you with questions, including suggested format for marketing your credentials. See the sample resume that follows for suggested format on the items that follow. Your resume will include:

- Your contact information
- A concise objective
- Your education
- Additional categories: relevant coursework, specialized skills, employment, experience, and so on

CONTACT INFORMATION

Your contact information should include your name, school address, home address, phone numbers, and e-mail address.

OBJECTIVE

A concise objective should be one to three lines long, beginning with your major, minor, type of work requested (e.g., new graduate, internship, or co-op position), and possible specifics.

You may want to have several versions of your resume with several different objective statements targeted to different types of employers; start with your broadest objective for general distribution. For example, if you are majoring in Civil Engineering and minoring in Environmental Engineering, you should create two resumes with the same content: one version with the objective more broadly to civil engineering and the second version with the objective specifically to environmental engineering. You would then send the second version to environmental engineering companies or agencies. The same applies to Mechanical Engineering (more general) and Aerospace Engineering (more specialized).

Your objective is what you can offer the employer— i.e., your education, experience, and possible areas for you to be valuable. Don't mention your personal desires— e.g., progressive company, growth opportunities, and so on. Focusing on the employer's needs is extremely important in selling, and that is what you are doing. To rephrase President John Kennedy's inaugural speech: "Ask not what your employer can do for you—ask what you can do for your employer!" The employer knows you have personal desires, but that discussion can come later. Initially, it's all about your ability to really focus on the employer's needs, and your ability to fill those needs. That will impress the employer and separate you from the crowd. It can get you the interview and, perhaps afterward, the job!

- ❖ A concise objective allows the employer to know at a glance whether you match the needs of the company. If so, they will read further—with interest.
- ❖ Use keywords in common, concrete language to describe your skills, so that a human resources representative or a computer resume scanner will understand your qualifications. Examples of keywords are:
  - Your engineering major and minor
  - Common terms describing functional experience (i.e., technician, clerk, mechanic, development, design, manufacturing, construction, sales, leadership, team)
  - Current buzzwords that best describe your objectives or skills

People with years of relevant experience, often begin their resume with a career summary of qualifications. In the case of most students without professional experience, the introductory objective is an accepted best approach.

EDUCATION

Next list your education, as shown in the sample resume. You should include your GPA if it is close to a 3.0 or above. Include both your overall GPA and that of your major. See more on listing GPA in the advice following the sample resume.

CATEGORIES

List categories that support your objective in bold and in descending order of importance related to your objective.

- ❖ Coursework relevant to your objective
- ❖ Specialized skills (computer, equipment/machinery, processes)
- ❖ Employment or experience
  - This is the most important, results-oriented section for experienced candidates, but also very important for students—even without technical experience.
  - Summarize employment experience and leadership, teamwork, or technical responsibility in school or any organization.
  - Arrange the list in reverse chronological order, i.e., most recent job or school club first.
  - Use bold print to name your employers/organizations or your positions. Don't bold-print both; be consistent in boldfacing one and capitalizing the other. Decide whether your job titles or the names of the organizations are more identifiable and meaningful in support of your objective and make them bold. State your position and dates of employment on the next line beneath the organization.
  - **Important:** Include action, results-oriented statements to explain what you did. See the sample resume for examples. This results-oriented approach positively separates you from most others who simply list duties without any mention of the results. Notice that the sample resume features bullets, followed by the past-tense of verbs. Once again, here is where you quantify results, when possible. Did you save money, make money, or create value for your employer or engineering club? How much? Express it in dollars or percentages if you can.
- ❖ Other common student resume categories could include:
  - Honors—Dean's list, Tau Beta Pi, or others
  - Engineering projects—brief summary plus results
  - Achievements—ones that make you stand out
  - Leadership—quantify accomplishments if possible
  - Activities—quantify results when appropriate
  - Special Skills—related to your objective
  - Languages—multi-language fluency
- ❖ References—"References available upon request" take up unnecessary space. Employers assume you have references.

SAMPLE RESUME

Putting all this together, let's look at a sample resume of Student J. Smith, a hypothetical Industrial Engineering student, seeking a summer internship.

STUDENT J. SMITH

123 College Address Town, State, 12345 (123) 444-4321 e-mail address	456 Permanent Address Town, State, 12543 (321) 555-1234

Objective	An industrial engineering summer internship in a manufacturing environment that will combine knowledge from technical coursework and interpersonal skills gained through employment.
Education	**Bachelor of Science in Industrial Engineering,** May 2012 UNIVERSITY OF MY STATE Overall GPA: 3.2/4.0, IE GPA: 3.4/4.0
Honors	Engineering Dean's List Name of Company Scholarship recipient
Relevant Coursework	Probability and Statistics, Engineering Economy, Operations Research, Ergonomics, Production Systems, Calculus
Employment	**Wegmans Food Market, Inc.** My Town, My State STORE CLERK, July 2009-Present • Monitored deli products to ensure compliance with health law requirements • Provided high quality customer service in a fast-paced environment • Maintained a clean and orderly work environment • Introduced more efficient system for stocking shelves **The Gap**, My Town, My State SALESPERSON, Summers, 2007 & 2008 • Served customers in pleasant and knowledgeable manner • Trained new employees and supervised staff of six in manager's absence • Won *Salesperson of the Month*, August, 2008
Computer Skills	Working knowledge of UNIX, Windows XP/Vista/7, MS Word, Excel, PowerPoint, Access, HTML, Maple, Matlab, C++
Activities	Institute of Industrial Engineers (IIE) Engineering Student Association UB School of Engineering and Applied Sciences Open House volunteer
Language Skills	Fluent in spoken and written Spanish

Comments on Student J. Smith's Sample Resume

Objective. Note that Student Smith immediately identified "an industrial engineering summer internship . . ." as the desired position. That's important because it will catch the eye of the person sorting resumes. Stating "a manufacturing environment" is fine if you have specifically targeted that as your desired area of contribution and experience. In that case, you would be sending your resume to manufacturing companies who would find your specific manufacturing interest very appealing.

Some people prefer not to list an objective, with the thought that they don't want to restrict themselves by specifying a desired work area. A brief statement of objective is desirable, since your goal catches the resume reviewer's eye and captures their interest if you are looking for the kind of job that they have to offer.

Education. You see that Student Smith listed an overall GPA of 3.2/4.0 and an IE GPA of 3.4/4.0. That complies with the general guideline of listing your GPA if it is 3.0 or higher. If your overall GPA is below 3.0 and your GPA in your major is above 3.0, list your major GPA. In fact, list both GPAs so not to be accused of misrepresentation later. The 3.0 guideline is just that—it's a guideline. Many experts believe that if your GPA is anywhere near a 3.0, you should list it so the resume reviewer does not presume that your GPA is closer to 2.0.

Employment or Experience. That means, once again, that you will list either employment or any other experience that will relate to a job that you are seeking. In the case of Student Smith, notice that the employers have been listed in bold type and the job titles have been capitalized, followed by dates of employment. Then duties and results have been set off with bullets followed in each case by the past tense of verbs. This allows the resume reviewer to quickly scan and digest what Student Smith has done with each employer.

Notice, for example, the last bulleted result under Wegmans Food Market, Inc.: "Introduced more efficient system for stocking shelves." If Student Smith had not been educated to write a results-oriented resume, Student Smith might have just listed a duty i.e., "stock shelves." If you were a resume screener, which statement would be more appealing to you? The answer should be obvious. The fact that Student Smith introduced a more efficient system for stocking shelves, in itself, might motivate the employer to interview that student and find out more about how Student Smith developed a more efficient system. In fact, if Student Smith could put a dollar or percentage value associated with that result, it could be even more appealing. In any case, that bulleted result is an example of what you will try to accomplish in your resume—summarizing evidence of specific positive accomplishments. Include engineering projects that support your objective; your engineering projects may be major selling points to catch an employer's eye. Don't go into any more detail than Student Smith did on the sample resume. You'll save that for the interview. Be sure to prepare for "how did you do it" questions; these are your main selling points and you will love to give brief summaries of your accomplishments during the interview.

YOUR COVER LETTER

The cover letter is the first thing the employer will see about you, so let's ask: how long does it take to make a first impression? Perhaps one or two seconds if the letter is really good or really bad. If it's good, you're still in the game. So, as with your resume, the cover letter needs to look professional without spelling or grammatical errors. Moreover, your cover letter is the gateway to your resume. It allows you to add some of your personality to the package, whereas your resume is a compilation of facts. Again, you're in sales!

The goal of a cover letter is having it and your resume read; an interview invitation is the desired result. I recommend two approaches for a cover letter: 1) generic for on-line employment Web sites and 2) customized for a specific company.

On-Line Cover Letter

If your university offers an engineering employment Web site, you want your cover letter and resume to be on it. You want all the exposure you can

GENERIC COVER LETTER GUIDELINE FOR CAREER CENTER WEB SITE

Your present address
City, State, Zip code

Dear Engineering Employer:

1st Paragraph—Briefly introduce yourself, your objective, and your areas where you can create value. You can begin by stating your engineering major and level, that you are applying for a co-op/intern/graduating BS/MS/PhD, and when you are available.

2nd Paragraph—Your aim now is to refer to your resume with some highlights that will motivate the employer to read your resume. Describe your specific qualifications you think would be of greatest interest to employers, keeping in mind the employers' needs. Mention any related experience or special training. Briefly summarize one to three key accomplishments. Tell the employer specifically what you have to offer and support your claim with evidence if possible.

3rd Paragraph—In your closing paragraph, ask for action—specifically an interview. Tell them that you look forward to the opportunity of meeting with them personally to discuss how you can be of value to their organization. Thank them for considering your application. Remember, selling the employer to interview you is the first step to selling the employer, during the interview, to extend a job offer to you. Don't go overboard, but don't be too timid either.

Sincerely,
Type your name here

get to employers. Your career center and/or co-op employment offices are a great resource for a large number of employers to do keyword searches for applications that meet their needs and for you to apply for job postings.

The document that includes your on-line cover letter and resume will be viewed by many employers, so the letter is general—that is, not customized and addressed to an individual and company. Center the letter so it occupies the middle of the page. See p. 50 for a formatting guideline for an on-line generic cover letter, describing the function of each paragraph.

Customized Cover Letter

If there is a contact person who you know, or know of, within a desired employer, you will want to customize your letter to that person and employer. A customized cover is similar to the generic letter, but with special difference in the fact that you know who you are writing to and are able to be more specific about why you're writing and the job for which you are applying. Personalize it as appropriate, but the content will often be the same basic information as in the generic letter.

CUSTOMIZED COVER LETTER GUIDELINE

Your present address
City, State, Zip Code
Date

Mr./Ms. Contact Name
Position Title (if known)
Organization Name
Address
City, State, Zip Code

Dear Mr./Ms. Contact:

1st Paragraph—Besides the personalized salutation to a specific person and company, the first paragraph begins differently. Some examples:

- I saw with great interest your job posting for a ________ engineer on the ________ University Career Center Web site.
- I am responding to your need for a ________ engineer that was advertised in the (date) *Daily Tribune*, (date) *Engineering News* (specify source)
- Mr./Ms. John/Jane Doe suggested that I contact you regarding your need for a ________ engineer.
- It was a pleasure meeting you at the ________ University Technical Career Fair on (date). Our conversation left me with enthusiasm to apply for your position of ________ engineer.

For a personal friend you can begin the salutation—Dear Sam/Samantha, I am following up on our conversation regarding (name of company's) need for a ________ engineer.

These sample first sentences allow the reader to connect you with why you're writing, how you got their name, and the position for which you are applying. The customized letter takes a bit more time, but it is well worth it from the standpoint of the probability of a positive response (or any response). The first sentence is the eye-catcher, particularly if you have met the person previously or you have a connection within the organization. Following the first sentence, you can use the information from your generic online cover letter to introduce yourself, your major, and any special qualification for this position.

2nd Paragraph—This paragraph can consist of the same information as your generic letter. The exception to this is that you want to, where possible, refer to the potential employer and open position by name and why you are specifically qualified for that job. As with the generic letter, the function of the second paragraph is to excite interest in reading your resume based on evidence of your qualifications and pertinent results. Again, it's the second paragraph that bridges to your resume and creates a desire to read further.

3rd Paragraph—This paragraph's function is the same as the generic on-line version. You are asking for an interview. Because it's a customized letter, you will name the organization and perhaps convey more specific information than in the generic letter. Thank the reader for reviewing your qualifications and express interest in meeting them personally to discuss how you can create value for their organization. In a customized letter, when appropriate, you could add a statement such as "I will call you next week to see if we can arrange an interview appointment." That "I will call you" statement is on the assertive side, but if you're comfortable with it and especially if you know the person you're sending it to, it could help you get the interview.

Chapter Summary: Takeoff Tips

The resume: your personal, concise, results-oriented, sales document. It will:

- ❖ Tell employers who you are and what you have to offer.
- ❖ Assist you in promoting your unique brand and marketing your credentials.
- ❖ Get you an interview—its main purpose. To obtain an interview, your resume will:
  - Be visually appealing
  - Be well organized and scannable
  - Be results oriented, with keyword evidence of your accomplishments
  - Emphasize your strengths and qualifications related to the employer's need
  - Serve as an interview outline for you to explain your qualifications

The cover letter: introduces your resume and you to the employer

- ❖ Can be generic, for your career center Web site posting
- ❖ Can be customized, for best results, to a specific person at a targeted employer
- ❖ The cover letter content generally includes:
  - First paragraph—introduces you and your objective to the employer
  - Second paragraph—encourages the employer to read your resume
  - Third paragraph—asks for the opportunity to be interviewed

Exercises

1. Write your resume, referring back to the narrative and sample as necessary. Use your judgment regarding the best format to market your objectives and skills.
2. Write a generic cover letter to be posted on-line for any employer.
3. Write a customized cover letter, to a specific person and employer.

Looking Ahead

Now, with your resume and cover letter completed you are well prepared to put your job search campaign into gear. You are ready to reach out to the right employers and the right functions. To connect with targeted employers you will want to utilize your university's career center, co-op office, or division of professional practice. Why? These offices are staffed with experts who are ready to help you with your career choice and obtain the best, suitable job. See what they have to offer and profit from using their services.

CHAPTER 6

Your University's Career Center

Your university has a career center. Use it. Typically, a university with an engineering school has an excellent career center and often a co-op/internship, experiential education, or professional practice program office. If you haven't used one of these career offices, please go there now. Meet them and see how they can assist you with your career focus, your employment issues, graduate school issues, or any other career matters. You will see that their career counselors and related services will be of great help to you. In fact, your career and co-op centers will most likely parallel and support material that we have covered in the last three chapters: self-analysis, market research of engineering opportunities, and writing your resume and cover letter. Your career center professionals will be glad to know you have done your homework on these subjects. Moreover, they can help you take your *Ready for Takeoff!* preparation and put it into action. How can they help?

PROFESSIONAL SERVICES

Career Planning, Job Search, or Co-Op Courses/Workshops

Career centers and co-op programs will often provide career planning, job search, or co-op courses/workshops for you. Enroll in these courses or workshops; you may be surprised at how much you will learn. Presenting yourself effectively to the right employer is crucial to your professional success and there is much useful information to help you obtain the right job. Career or co-op courses offered by your university, along with this book, can substantially enhance your opportunity to get where you want to go.

Individual Counseling

A career services counselor can help you by:

- ❖ Reviewing your self-analysis and market research of engineering opportunities. A counselor can study your conclusions and perhaps bring up additional ideas worth pursuing. They may also be able to provide more resources to use.
- ❖ Reviewing your resume and cover letter. Your counselor may help edit and polish them. Counselors also provide resume and cover letter content advice. Another set of professional eyes on your resume and cover letter will ensure there are no errors in spelling or grammar. You'll be confident that your resume and letter are first-class before an employer sees them.
- ❖ Helping with other career questions, including employment search tips tailored to your needs. As you discuss your qualifications and objectives, your counselor may have employers that come to mind to whom you may want to apply—perhaps employers that you would never have known about were it not for that discussion.

Career Center/Co-Op Program Web Site

- ❖ See all the services offered. More than you thought? Plan to use them.
- ❖ Upload your resume and cover letter, along with a listing of courses and grades. Career center Web sites generally convert Word documents to a PDF file. Your career center or co-op office will have instructions for uploading and using the specific career service software used by your university. As explained in Chapter 5, employers use keywords and other search criteria to select on-line candidates of interest. Keywords are standard words that describe your major, computer skills, and common terms to describe your functional experience. Recall that your resume will stand out by using common keywords that an employer would seek that are related to your technical, leadership, teamwork, and communication skills. Employers will commonly search for your qualifications, using such keywords, and may contact you directly or through the career center. They are looking for a blend of technical and soft skills, so use standard words to describe your qualifications in such categories.
- ❖ See on-line posted positions from employers and apply to those fitting your interests and qualifications. You can do the same as above with keywords to search for employers whose needs match your skills. Your career center or co-op program office will have instructions on this, as they did for uploading your resume and cover letter.
- ❖ Other features: career center Web sites offer other features such as resume/cover letter writing software tools, self-assessment tools, employer databases with links to employer Web sites, and more.

Career Center Library

Career Centers may have a library with a variety of information concerning employers, careers, and job searching. Typically there will be a selection of materials to explore jobs, careers, and graduate schools. The library is very useful as you research specific areas of opportunity.

Practice Interviews

Career centers will sometimes offer practice interviews. You may be able to be videotaped in a practice interview with a career counselor, then watch the video recording to be coached on your interviewing skills including anything that needs improvement. Some career offices have on-line practice interviewing software as well. Musicians and athletes wouldn't think of going into a performance without practicing, but most people don't think of practicing and being coached before a job interview. That's unfortunate for most, given the importance of a job interview. If your career center has a practice interviewing service to offer and you use it, you will be ahead in the game. You will know what to expect, be more confident, and increase your chances of a positive outcome during the actual interview.

Employer Campus Interviews

This is the service most students know about. Employers will visit your campus to interview students for their engineering employment needs—co-op, internship, or new graduates at the BS, MS, or PhD level.

Procedure. Go to your career center and/or co-op office or its Web site to see what campus interviewing process is being used at your university. It is common that the career office will post the dates employers will visit, including job descriptions, required qualifications, and deadlines to submit a resume for pre-screening or to sign up for an interview.

Pre-Interview Information Sessions. Many employers will conduct an information session the evening before interviews. If you're being interviewed the next day, you need to go to the session, even if you have researched the employer on the Internet. The employer will explain details of their company and their recruitment needs. Attending an information session before interviewing gives you three advantages:

1. A good first-impression with the employer: you are interested in the company.
2. You'll be prepared to tell the interviewer the next day how you're qualified.
3. You'll be able to use more interview time selling your skills to the recruiter for a targeted need, rather than having the recruiter repeat company information that was explained at the information session.

Take Action Early. Employers will often announce their openings and arrive on campus as early as September to interview students for projected openings in May or June of the following year. That early? Most students aren't ready with their updated resumes—including last summer's job experience—having it uploaded on the career center or co-op Web site by the start of the fall semester. Too many students don't catch on to the "early action requirement" until the following spring, when they sadly learn that they missed out. If you want to make the best of campus recruiting, you need to get an early start, and you'll probably be very thankful you did. Many employers are looking for the students who take the early-start initiative. They also are often anxious to get employment decisions completed as soon as possible, even by the end of December. Wouldn't you like to have a great offer nailed down by New Year's Day? Then you wouldn't have to be stressed over getting a job during spring semester.

Job Fairs and Graduate School Fairs

You won't want to miss this opportunity to meet with many employers from across the country and, perhaps, representatives of graduate admissions staff from a number of universities. Your career center organizes these events to allow large-scale exposure of students to employers and employers to students. These fairs are generally set up in an area where employers and graduate school staff have assigned tables and displays about their organizations along with recruitment materials. At each table there will usually be two or more representatives to talk with. The employer representatives will evaluate interested students and make recommendations as to whether you are qualified or not.

Your career office may publicize which organizations are attending the fairs, so you can prepare in advance and choose those employers or graduate schools that you are interested in pursuing. Check out their Web sites and you'll be among the most prepared to know the employers' technology, markets, and how you could contribute positively to their organizational objectives. Having done your homework, arrive early at the fair dressed in business attire, as suggested by your career office, with an ample supply of resumes to give to representatives of organizations that interest you. Their tables will have brochures of employment opportunities; pick one up and read it while you are standing in line to talk with a representative. One more thing: before the job fair, prepare and somewhat memorize an elevator speech, or self-commercial of 20–30 seconds, that you will give to employers or graduate school representatives who you meet. An elevator speech is a quick pitch that you could give to a boss on an elevator before reaching the 10th floor.

You're on. Hand career fair representatives your resume; shake hands firmly; look them in the eye; smile; introduce yourself; tell them your major; whether you seek full-time, internship, or co-op employment (or grad school admission); and give your brief elevator speech of why you're coming to their table. Let the representatives know that you researched their organization on the Internet and/or in their recruiting brochure and have some preliminary idea of where you could fit in and how you could create value for them. Your preparation includes your technical and soft skills related to their specific employment needs or graduate school programs. The recruiters might then explain opportunities that would suit you. See whether the representatives indicate positive interest, and get their business cards so you know who you talked with for possible follow-up. Employers will often make a notation on the back of your resume that they were impressed with you and recommend that their organization interview you further on-campus or at their location.

Meet with as many employers or graduate school representatives as possible during the hours of the job fair. These fairs provide great practice meeting employers, exploring career options, and selling yourself. In two or three hours you will gain much career opportunity knowledge and self-confidence.

Meet a Mentor

Would you like to talk to an experienced engineer in your area of interest? Some career centers may have a database of experienced alumni, including engineers from various disciplines and functional career paths, who want to share their experiences and are interested in helping students. Many schools also offer a mentor-connection service through their alumni associations. Meeting a mentor in person, by phone, or via e-mail can lead to valuable wisdom and advice from your mentor's perspective. Often there will only be one meeting, so prepare to make the most of your time with the mentor volunteer. Remember, you're asking the mentor to share information with you, not give you a job. This can be the start of an ongoing relationship; you may want to connect with the mentor throughout your career. Networking is extremely important in your job search, and meeting a mentor through your career center is one networking approach. We will cover more on networking in Chapter 7.

CONCLUSION

The use of the services of your career center will be very worthwhile. It will put you ahead of the crowd, save you time, and enhance your chance of success.

Chapter Summary: Takeoff Tips

Your career center can help you by offering professional services such as:

- ❖ Career planning or co-op preparation workshops
- ❖ Individual counseling
- ❖ Career center/co-op program Web site
- ❖ Career center library
- ❖ Practice interviews

- Employer campus interviews
- Job fairs and graduate school fairs
- Meet a mentor

Exercises

1. Which offerings of your career center do you use?
2. Which service do you plan to use more of? How? When?
3. What can you do so that your career center may be most helpful to you?

Looking Ahead

Chapter 7 will give you complete information on how to plan and conduct your job search campaign. You will discover a number of resources and actions you can take, many of which are seldom used but are extremely effective. Act on the contents of Chapter 7 and obtain interviews with the right employers.

Organizing and Executing Your Job Search

Now, let's put it all together to organize and execute your job search. At this point you have done a lot. You have worked through logical steps toward a good career start. Specifically, you have done your self-analysis in Chapter 3 and reviewed the types of engineering opportunities in Chapter 4 that could suit you. You have also used Chapter 5 to write a resume and basic cover letter. Hopefully, you are connecting with your career center or co-op office and taking advantage of their offerings described in Chapter 6. This chapter will continue the logical progression of steps toward your career launch. Now it's time to suggest organization and action steps that will bring you to the right employer. Your preparation done so far will pay off, so this should be fun.

THE LAUNCH

We're now going to outline things you should consider doing. Why? So you can get the job you really want while still keeping your head above water with your engineering coursework. Here are action steps you should take.

ORGANIZE JOB SEARCH MATERIALS

That includes making a file of your resume and copies of all the potential opportunities, correspondence, and so on that lie ahead in your campaign. Have it near your phone and computer so you can access it readily when you need it.

NETWORKING

That's the one-word best advice you will hear from literally all job search experts. Networking means that you know someone who can hire you, or you know someone who knows someone who can hire you. That reality, and realization, is hugely important. It means that you have friends, relatives, acquaintances, professors, former employers—people you know or people they know—who you can approach for an information interview or an actual job interview. Most people are on good terms with their parents, relatives, and friends. These people are a terrific resource, and they'd like to help you. The unfortunate thing for most job hunters is that they under-utilize this great network resource. Most people don't sufficiently spread the word that they're job hunting and they don't hand out resumes that could then be passed on to a target employer or information interviewer. For your sake, consider being one of the minority of job seekers who networks.

Specifically, How Do You Network?

1. *Print a Large Number of Resumes.* Print at least 100 to start, so without hesitation, you can hand them out to anyone who is in a position to talk with you about your professional goals. That person could help you directly or could be a person who knows what you are looking for and can pass your resume on to someone who can talk with you directly about your career.
2. *Business Cards.* Have you thought of having your personal business cards printed? No? Well, you're not alone. I'd estimate that fewer than 2% of student job seekers have business cards to hand out to people they meet. The beauty of business cards is that you can carry them everywhere in your wallet, purse, or pocket when you don't have your resume with you, such as when you're at a party or community event. Business professionals who are on the ball *always* carry business cards. They do that because they never know when they might meet a potential customer or someone they want to have in their network—someone they want to have remember their name and contact information. They give business cards and obtain business cards to promote positive action and widen their network. Make no mistake—even the most successfully employed professionals (especially the most successful) know that they will be networking until, even after, they retire. I tell my students I carry an ample number of business cards to business functions, parties, weddings, bar mitzvahs, in my ski jacket, in the pocket of my sailing shorts, anywhere I go. I constantly meet people who want, or know someone who wants, to hire an engineering graduate or co-op employee. Carrying business cards works!

   If you hand out and receive business cards, you, as a student, will immediately look professional and stand out from the crowd. Moreover, transmitting and receiving business cards will be very helpful information for your networking process.

   When you don't have your resume with you, request a business card from a new network contact to provide you with their name and address so that you can send your resume, either by e-mail attachment using your cover letter as the basis of the message or by postal mail with a cover letter. Your card reminds the person of your name, your degree, your conversation, your goal, and to expect a resume.

Business cards are inexpensive and can be obtained at your campus print shop or a local printing or office supply store. Plan to order at least 100 cards. For a few more dollars you might get 250 or 500. I suggest you check on your campus to see whether the campus print shop can provide you with your university logo and name across the top. Ask the printer to arrange the business card as follows:

Your Name
BS Mechanical Engineering, May 201X
Name of University (if not printed with university logo on top)
XXX Hall (address)
City, State, Zip
Telephone: (987) 123-4567
Fax: (987) 123-4568 (if you have one)
E-mail: yourname@internetprovider.com

3. ***Your 30-Second Commercial.*** Before you start meeting people with whom to discuss employment, have your elevator speech or 30-second commercial prepared as described in Chapter 6. Then when you do meet someone, introduce yourself, give your elevator speech that includes your degree and area of job interest, and hand the network-contact your resume or business card. That combination goes a long way toward a positive result. Your 30-second commercial and resume informs the network person of who you are, your qualifications, and what you want. That person is now prepared to give you any information that can lead to positive action toward your employment.
4. ***List Names of People or Organizations You Want To Contact.*** For purposes of being organized, you may wish to have a database of employers, relatives/friends, professors, mentors, or engineering professional associations who you want to talk with, along with their contact information. Your list will grow immediately as soon as you begin to network, so it is best to divide your database into different categories. Expect that your growing list of contacts will bring you ever closer to the kind of job you want with the employer best suited to your skills, interests, values, and objectives. You should consider:
   - Identifying target employers that could meet your objective
   - Writing statements summarizing how you could meet their needs

   If you have target employers and an idea of how you can contribute to them, it will help you in networking discussions. It will be effective to have this information organized and ready for conversation with your network contacts.

   If you don't have that information now, that's OK too, since networking discussions and Internet search will help you determine target employers and how to match your skills and interests with their needs.

5. ***Start Networking Discussions.*** Warm-up by beginning with people you feel comfortable with. Soon you will feel confident to talk to anyone. Show a network person your resume. "Decompress," that is, let that person know you're after information, not a job offer from them; give the 30-second commercial and see what they have to say. They may give you some feedback or suggestions about your resume, especially if you invite them to. They may suggest you talk with another person or organization. They may want a few more copies of your resume to pass along to whomever. Just be sure you don't "grab them by the sleeve" and ask them for a job! It would put your network person on the spot by asking for employment. That comes at the end of the next chapter on interviewing. "Closing," or asking for a job offer, is not done in this networking stage. Decompress is the advice by professional employment coaches during information interviewing. When you decompress, it means that your network person understands you're just after some quick information or employment leads. It's best to come right out and say so.

   The point of networking is to get more information about people and organizations that might hire you and to develop a more refined idea of just where you belong and what you would do best. Review your self-analysis from Chapter 3 so you can easily summarize your skills and interests. Likewise, review your assessment of engineering opportunities from Chapter 4 so you can let your contacts know what generally interests you. This combined information from Chapters 3 and 4 will allow your network contacts to ask you questions and give relevant information toward your goals. When you conclude each information interview, try to get the names of several other people or organizations you will talk with next. That way your network continually expands. So, in summary, networking involves a sequence of contacts bringing you ever closer to organizations and people who can hire you:

   - Your initial contacts—people who you know
   - Referred contacts—people your initial contact knows who might hire you
   - Final target contacts—people who could hire you

6. ***Summarize the Results of Each Network Discussion.*** You could have many discussions. However good your memory, don't think you'll remember all of the important points and actions to be taken; put your summary on paper or in your computer. Don't hesitate to take notes during your discussion; in fact, please do. Obviously, if your network person is giving you specific names, phone numbers, e-mail addresses, or feedback on your resume, you want to be prepared to take notes. Your network contact will then know that you are professional and take the conversation seriously. That makes him or her even more motivated to help you. If you can't make notes during the conversation, be sure to do so immediately afterward, including action steps for yourself or action you want your network contact to take.

7. *Thank Your Network Contacts.* As a professional courtesy, form the habit of sending e-mails or hand written notes thanking each person who took the time to help you. That's an important success habit throughout life. Now, as a job hunter, it will serve you even more. Specifically, your individualized thank you notes will remind your network contacts to take some action on your behalf if they haven't yet done so. Express what you learned from them and remind them of any action conclusion you and they agreed on. That will be helpful to your network contact and, most of all, to you.
8. *Sorting the Advice.* You decide what advice is especially useful to you. In networking, you may receive a fairly wide range of opinions—some may be conflicting opinions. That's great, since it can give you food for thought about the pluses and minuses of various alternatives. Keep thinking and moving ahead, and don't be stalled by "paralysis of analysis." Also, don't be overly influenced by a persuasive person with different values than yours. Decisions about your future are your own. You've done your homework and you're reaching out, refining your ideas about where you belong and how to get there. You'll get an increasingly better idea what's right for you as you proceed through the process. In summary, through networking, you will be able to identify and really close in on employers and key contacts who could hire you.

INTERNET SEARCH

Your career office and network provide extremely valuable means to employment. By all means, use your career office and network! We cannot, however, leave out the Internet as a wealth of employment information at your fingertips. The Internet can provide you with a lot of detailed information that will enable you to connect with specific organizations, people, and jobs. You should do research on the Internet before, or as part of, your networking discussions.

If you Google-search "engineering jobs" you will be able to connect with some great sites through which you can aim and hit a specific target. In fact, you could get carried away; there are, at this writing, no fewer than 65 screens listing "engineering jobs" Web sites, with about 10 site-listings per screen. The screens progress from large general sites to, toward the end, those with more specific job locations or functions.

Large, General Engineering Sites

Let's look at two as examples:

- www.engineering.careerbuilder.com
- www.engineerjobs.com

These sites are toward the top of the Google listing and are two of the largest Internet engineering employment sites. Sites such as these allow you to browse for jobs in any state, city, or town. You can also search by engineering job category and engineering discipline. Take a look at them and notice the ability to get your customized listing of job opportunities, based on your selection of keywords, discipline, location, or other search options. You can review specific job descriptions and apply online.

Professional Engineering Society Sites

If your university has a student chapter of an engineering society for your major, it's a good idea to become a member. Student society membership is low-cost and gives you opportunities for leadership, technical projects, networking, and, in this case, Web site use for job search in your major. Here are three engineering society Web sites:

- www.aiche.org/Students/Careers/index.aspx for chemical engineers
- www.ieee.org/web/careers/home/index.html for electrical engineers
- www.asme.org/jobs/ for mechanical engineers

These and professional societies of other majors' sites provide you with the opportunity to connect with your professional society. Often, these sites give you the chance to select a listing of jobs and provide information to send in your application.

Social Networking Services

Internet social networking has become very popular. It has become an important means of connecting with many friends and potential employers. Here are two major services:

- ***LinkedIn.*** This is a professional networking service. As such, it is useful for networking on a business level. If you are a registered user of LinkedIn you can request, establish, and maintain "connections" with people you know or who they can recommend to aid you in your employment search. Obtain more information by visiting www.linkedin.com. People in your connection network may help you by possibly suggesting areas of potential opportunity, including contacts within a targeted organization. This means you might have a LinkedIn connection who could introduce you to a hiring manager. Employers can also use LinkedIn to post available positions.
- ***Facebook.*** Using Facebook, you can set up a personal profile page and connect with other "friends" for the purpose of exchanging information. Social networking allows you to widen your personal network. You can set your profile on "personal" or "public" that has privacy implications. Your profile page includes a "wall" on which friends post messages and photos. Your privacy settings control who might see your wall. Therefore, be aware that potential employers might read your profile page, including your wall. Would you like an employer to visit you on Facebook? Does your profile and wall represent you well as a prospective employee?

There is an option to join networks specific to location, school, and other criteria. For more information visit www.facebook.com.

- *Other Social Networking Sites.* You can Google "social networking sites list" and review dozens of other resources, such as Twitter and MySpace, by clicking on the descriptions of their services.

STAY ORGANIZED

Using your career/co-op office, networking, and the Internet, there are multiple approaches for you to obtain a job offer from the best place to start your career. There are many people and employers you will be contacting. Here are some ways to keep track of it all.

- *Networking Contact Lists.* Make a contact listing file. This will include list headings of people and organizations with whom you will network. Having the listings categorized increases your efficiency by helping you list and later find names of important contacts. The group of lists will grow as your campaign progresses.
- *Employer Contact Summaries.* Summarize information about each employer you are contacting and action taken as you proceed with them. List names, key contact information, and a log of action taken as you go along. This lets you stay organized with action that you and each employer have taken, and reminds you of next steps.
- *Make a File System.* Include your resume, cover letters, along with the above-suggested networking lists and employer contact summary sheets in your file system. Your file system can be electronic, paper, or a mixture of both. Proper filing enables you to locate important documents and avoids the stress of not being able to find information when you need it. A spreadsheet format is an excellent electronic tool. Your school may have an electronic database system available for you to organize everything.
- *Monthly Calendar.* Buy an inexpensive monthly calendar if you don't already have one. Record any important dates in your job search campaign for that month. That includes career fairs, interviews, or follow-up dates. Your calendar allows you to organize your month and see all entries at a glance so that you don't miss an important day or time for you to take job search action.

RESEARCH SPECIFIC EMPLOYERS

Now, research specific employers within your market niche before interviewing.You are identifying and researching employers where there is most likely to be a mutual fit for you and their needs.

- Study employers that interest you and that could use your talent. Find out about their products, their technology, and their customers online or at your career office. If possible, evaluate specific job openings. How could you fit in as a contributor? Are you excited about the possibility of employment with this organization?

- ❖ Are any of these employers recruiting on your campus?
- ❖ Will any of these employers be at a career fair on your campus?
- ❖ Do you have a network contact who could connect you with a representative of that employer?
- ❖ If none of the above, you can apply via their Web site. Often employer Web sites have detailed listings of their job openings, so you can plan your discussion of how you can meet their specific needs.

FINAL ACTION STEPS

Before being interviewed, you will want to do the following:

Identify and Pursue Your Prime Employer Prospects

You may have done a lot of employer research, from general to specific. Now it is time to select the most promising employers and ask for interviews. That means submitting your resume and requesting an interview from employers who are visiting your campus career center. Also, send e-mails or letters with your resume to specific companies that you located through the career center, networking, Internet search, or any other means. Research the company so you and they will know you can be a good fit for their needs.

Customize your interview request cover letter and resume to personally let the employer know of your specific interest in them and why you are qualified. You prepared a general cover letter in Chapter 5, now customize that letter to a desired employer by addressing it to the target contact by name and title—perhaps someone your network person referred you to—and the name and address of the employer. Specifically state in the first paragraph why you are applying and include who referred you. You can use your standard second paragraph, but look for an opportunity to connect any of your specific skills to their specific needs. Repeating Chapter 5 advice, your last paragraph will ask for an interview so you can discuss how you can add value to their organization; it's important to approach an employer not as a person needing a job. Rather, you are a person who knows how your qualifications connect with their needs and are passionately interested in creating value—creating profit—for them.

Your message to an employer can be by e-mail cover letter with resume attached or postal mail on good stationery. Proofread it before sending to ensure no grammar, punctuation, or spelling errors and keep electronic or paper copies of correspondence filed so you can access them quickly—like when the employer calls.

Your Wardrobe and Accessories

First impressions are very important, and you will want to make a good one by looking professional. That means having a conservative interview suit that fits you well with a nice shirt or blouse and tasteful accessories—a tie or a scarf, jewelry, and so on.

Have a watch, a pen, and appropriate shoes; remove any visible body piercing before the interview. Another key accessory is a professional looking folder that encloses copies of your resume, unofficial academic transcript, names and contact information of your references, a printout of the employer's job posting, and a pad of paper for taking notes.

References

Line up three or four people who know you and can give you a positive recommendation if an employer calls them following the interview. Ideally, it will include one or more previous employment supervisors, a professor, and a personal character reference—someone who has known you for a long time and can attest that you are a good, law abiding citizen with solid values. It is important to ask prospective reference people if they agree to be your reference and whether they would give you a positive recommendation. If they hesitate, or say no, move on to your next prospect. Give your reference person a copy of your resume and briefly summarize your objectives and why you feel qualified and motivated to achieve them. This preparation of your references can make all the difference. Now they've agreed and are prepared to help you; they are aware of your qualifications and objectives and can speak specifically, intelligently, and positively about you as an employment candidate. Properly selected and prepared, your reference person is likely to do a good selling job to an employer on your behalf. Doing these steps with references puts you ahead. Why? It's amazing how many applicants list people as references without even contacting them first. First of all, that's impolite. Moreover, it's no surprise in such cases when the reference, on receiving a phone call from an employer, struggles to remember the candidate, can't describe specific qualifications, or may even give a poor recommendation. An unprepared reference situation is unfortunate and avoidable if the right steps are taken.

Voice Mail/Answering Machine

You should have a telephone message system when you are unable to answer the phone directly. During a job search you are going to receive and leave important messages that will exchange key information and make important first impressions. So remember:

1. *Receiving a Message.* Program your voice mail answering message to say something like "This is (your first and last names). Please leave a message and I'll call you back. Thank you." Clearly stating your name and politely requesting a message is important. The humorous or offbeat message from your freshman year is not one you want a serious employer to hear as their first impression of you!
2. *Leaving a Message.* When leaving a return message for an employer, it is important to start by slowly and clearly stating your first and last name and telephone number. Then concisely state your message, letting

the employer know of your availability for an interview or answering whatever was asked. You might begin with a reminder, "I'm responding to your message; we met during the career fair at ________ University on (date)." Conclude your message with, "Again this is (your name), (phone number). I look forward to the interview. Thank you for your interest." The reason I emphasize a repeated slow, clear pronunciation of your name and phone number is that a pet peeve of many employers is getting a message from a student where the pronunciation of the name is a blur and the phone number, often at the end of a long message, is stated much faster than the employer can write. Don't expect an employer to replay your message multiple times to understand it.

CONCLUSION

Now, you're on course to meet some of your selected employers. Everything you have done so far gives you the opportunity for the big payoff. You are going to be interviewed, and you are going to be ready.

Chapter Summary: Takeoff Tips

Organize and execute your job search.

- ❖ Organize your job search materials and stay organized.
- ❖ Network with contacts who lead you to employers who could hire you.
    - Print a large number of resumes and business cards
    - List names of people or organizations you want to contact
    - Use the Internet for job searches and professional engineering sites
    - Use LinkedIn and appropriate social networking sites
    - Connect with initial contacts, referred contacts, and final target contacts
- ❖ Identify and pursue your prime employer prospects.
- ❖ Research targeted employers before interviewing.
- ❖ Prepare for interviews.
    - Know how you can contribute value to a targeted employer.
    - Your wardrobe and accessories.
    - References.
    - Voice mail/answering machine.

Exercises

1. Explain how you have organized your job search materials.
2. Explain why you will network. How? With whom? Describe any of your networking results so far, including possible use of your resume and business cards.

3. Display your personal business card to be printed—(model, page 63)
4. Explain how you plan to or have used the Internet for your job search. What sites? Results?
5. Explain how you have, or will have, researched specific employers before interviewing. (Perhaps through the Internet, Career Center, your network, and publications.)
6. Identify your prime employer prospects.
7. Ensure you are ready for interviews—with preparation of your self-assessment, resume, wardrobe, references, voice mail/answering machine, and employer information.

Looking Ahead

The next chapter will describe all the details about effective interviewing, geared to get your job offers. Most job candidates are not sufficiently prepared to be interviewed. Study Chapter 8 and you will be better prepared and more confident during interviews. In fact, interviews can be enjoyable.

CHAPTER 8

The Successful Interview

Congratulations. You've got the interview. You are about to be interviewed because of your good resume, your market research of the right employer, and perhaps some network contact that led you to the right place. You're going to meet an employer who thinks that you could be the right candidate for the job. Now the employer is going to invest valuable time with you to see if they can offer you a job.

WHY SHOULD I HIRE YOU?

The main question you will answer affirmatively for the employer is "Why should I hire you as opposed to someone else?" Consider:

Assumption #1. The employer has a business need to hire a person.

Assumption #2. The employer wants to hire the best person to fill the job and will pay a competitive salary to do that. The employer is the customer.

Assumption #3. If you have been following the job search process described in the previous chapters and act on the contents of this chapter, you can present yourself as the most qualified candidate with the best-presented evidence of what the hiring employer seeks. That can lead to a job offer!

Assumption #4. It's not necessarily the most qualified candidate who gets the job—it's the person who appears to be the most qualified for that job with that employer. That's reality! The employer can only process evidence that they see, and you're in the business of honestly selling your strengths to win the apparent match of what the employer is looking for.

WHAT'S THE EMPLOYER LOOKING FOR?

Once again, the underlying question is "Why should I hire you?" The most fundamental things an employer seeks in an engineer, or anyone, are provable:

- ❖ ***Competence.*** Your ability to do the job. "Can do."
- ❖ ***Motivation.*** Your interest and passion to join the organization and do the job. "Will do."
- ❖ ***Personal Qualities.*** Your personality and other qualities enabling you to fit well in the organization, including working on teams and with customers. "Will get along well."

The Employer Wants You To Succeed in the Interview

That's an important, optimistic mind-set for you before going into the interview, and it should help boost your confidence. The employer hopes that you will be the right candidate. Why? Because they were impressed with your resume and hopes that you will live up to their pre-interview expectations. So, don't disappoint. Be prepared. Interview well, so that:

- ❖ They can offer you the job and hopefully get an acceptance.
- ❖ They will fill the job and won't have to waste time interviewing others.

Look for the Win-Win

Let's face it: both you and the employer have important goals that are similar:

- ❖ You want a job that fits your skills, interests, and objectives.
- ❖ The employer wants a candidate with the skills, motivation, and the right personal qualities to be a value-creating asset to the organization.
- ❖ You and the employer are investing your time in conversation to see whether this is the right match. It's a bit like looking for the right life partner—you both need to be satisfied that there will be success in a very important relationship.

TAKING ACTION DURING THE INTERVIEW TO GET THE JOB

How do I do it? What do I do?

- ❖ First, remind yourself of Important Reality #3 from Chapter 2: you are a salesperson. You are selling yourself to the employer. You are selling yourself, with evidence, that you are the best qualified candidate to fit the employer's specific needs.
- ❖ Second, review your self-assessment that you completed in Chapter 3. That will give you the information about yourself that you need to sell to the employer.
- ❖ Third, review your study of Chapter 4 and your market research to see that your choices of engineering opportunities align with the organization

that is about to interview you. Define your fit. What can you contribute? Do you have a passion for their work?

- Repeating advice from Chapter 4, study potential employers and job functions to see that you are aiming for the right potential targets. Confirm this by taking interviews. You don't want "paralysis of analysis" by hesitating to take interviews. Nor do you want to go off in all directions without thought, wasting the time of both you and employers.

Focus

Focus on what you can do for the employer, not on what the employer can do for you. Employer-focused contribution was emphasized in Chapter 5, but it is worth reiterating now. It is crucially important during the interview that you emphasize what you can contribute to the employer, rather than vice versa—meeting your personal needs. The focus on what you can contribute, rather than what you can receive, is a basic success outlook.

Approach

It is crucial to approach the interview with a clear understanding of:

- The employer's specific job requirements.
- How you can, with competence and enthusiasm, fulfill those requirements.

That's essential, logical, and simple to state—but most people don't do it. Why? Because they do not understand the process of getting the right job and/or are too lazy to put in the effort for success. Conveying how you can contribute is to your competitive advantage!

Prediction. If you are armed with the most specific information about what the company needs, alongside how you can specifically, competently, and enthusiastically meet that need, you can be the apparently best qualified candidate. That can mean you're on your way to a job offer. At every opportunity during the interview conversation, connect your skills, motivation, and personal qualities with their job requirements. Most candidates will not do that nearly as well as you—so be prepared and confident that the interview will go well.

C THE SUCCESSFUL INTERVIEW

Here is an outline and explanation of the steps for a successful interview.

Prepare

Know Yourself. Who are you? Know your relevant skills, interests, goals, and values as explained previously. Practice efficiently explaining your strengths and academic or job experience listed in your resume in a brief, organized way. Be prepared to enthusiastically show specific examples of your accomplishments. If you're able to quantify accomplishments, that is especially good. Grades are an example of accomplishment—particularly if you can show that you were able to apply theoretical concepts in the laboratory. Other examples: if you saved an employer money, how much in dollars or percentages? Specifying or quantifying accomplishments can apply to a school club or any organization as well. If you were a leader in an engineering club, did you increase membership by a certain percentage? Did your club win an intercollegiate technical competition?

Every engineering student has strengths that can be valuable to the marketplace, regardless of your academic record or experience. Be assured that there is an employer that is right for you—you just need to find an employer that is aligned with your values, strengths, and interests and present yourself in the right way, as explained in this chapter.

In short, know and be able to explain the factual evidence of your qualifications. Moreover, psych yourself up to present yourself as a motivated, personable individual who can work well on a team or as a leader. Employers are looking for engineers with a blend of technical qualification, motivation, and personality/communication skills to be all-around contributors within their organizations.

Know the Employer. Who are they? Before going into the interview, you will appear to be a best qualified candidate by not only knowing yourself, but with whom you are speaking. You will research the company, their products, their technology, their history, and their reputation. You will especially want to know the details of their job opening and how you can create specific value for that job and the organization.

Your ability to effectively connect your skills with the employer's needs will separate you positively from the rest of the crowd. Prepare yourself with intelligent questions that relate to job responsibilities, rather than personal issues such as benefit plans. If the employer is interested in you, they will tell you about benefits.

Know What Salary You Are Worth. Prior to the interview, it is important to know what you are worth in the marketplace. Why? You may be asked for your salary expectation during an interview. Moreover, when you receive an offer, you need to know if it is competitive. There is abundant information online about this. Pick your Web site. One that is popular is reached by Googling "salary wizard." That will bring you to a number of sites including Salary Calculator and CollegeGrad.com. Enter your engineering major at the entry-level and a zip code for a normal curve of salaries available in your discipline and your geographic area of choice. Check out the various sites to know your market value. The National Association of Colleges and Employers (NACE) publishes a salary survey of the previous year's beginning offers. Your career services office should have a copy of that report. It includes engineering offers within all majors at the bachelor, master, and doctorate levels. Within each major, the most recent average offer is listed along with offers at the 25th, 50th, and 75th percentiles. Generally, expect large corporations to offer a higher starting salary versus smaller companies. Companies pay extra for experience, including co-op, high grades, and high-potential leadership qualities. How much extra? Additional compensation varies, based your unique package of skills, the company's salary policy, their need for you, and their desire to get you on their payroll.

The Big Day—Before the Actual Site Interview

Be aware from the start:

- Dress to impress. Dress conservatively. A dark suit, such as navy blue, for men and a skirt or pants for women is commonplace for interviews.
- If you don't know the location of the interview site, use MapQuest or even do a dry run to ensure you know where you're going and how long it will take to get there.
- Early is on time; on time is late. Arrive 10 minutes before the interview. If you're late, it's over.
- From the moment you enter the employer's building, be aware that you are being evaluated. Be pleasant and polite when you introduce yourself to the receptionist, announcing your name and that you are here for an interview with ________. It is not unusual for interviewers to ask receptionists their opinion of you and how you behaved to anyone else in the lobby, including the maintenance staff. Companies are interested in knowing how people act when their guard is down and they don't feel the need to be impressive.
- Complete any paperwork, including application forms you are asked to do.
- Bring several copies of your resume, unofficial transcript, and list of references. Enclose them in a nice-looking briefcase or bound pad in which you can take notes. You could bring an impressive project report, especially if it relates to the employer's business or technical need.
- You should have a listing of at least three references: perhaps a professor, a previous employer, and a personal character reference who has known you for many years. List them on one sheet entitled "References for (your name)" and include name, title, address, phone number, and e-mail address for each reference person. Be sure you have asked them if you can use them as a reference and if they can give you a favorable reference. Provide your references with your resume and an explanation of your objectives and what you have to offer.
- These items may appear to be obvious. But you need to realize that many candidates flunk out on any of the above points before the interview, whether it's attire, arrival time, the handling of references, or courtesy to the receptionist. Fortunately, you will pass these first-round tests since you are informed, professional, and have common sense.

The Actual Interview—Summary

The Greeting. How long does it take to make a first impression? It happens in a matter of seconds, perhaps even a split second. You are well-dressed and greet the interviewer with a firm handshake, a smile, and a good eye contact.

Already, you are impressing the interviewer. You have impressed the interviewer as opposed to the previous candidate who handed the interviewer a limp handshake, failed to smile, had poor eye contact, or generally looked uncomfortable. That candidate may have flunked during the first few seconds of the interview. But that's not you—you're prepared and will continue to impress the interviewer.

Your Style. You will be sincere and enthusiastic during the interview, especially in explaining how your skills may match the employer's needs and providing concrete examples. The interviewer is in charge, asking you questions for your honest response. You are having a friendly business conversation.

Keep Your Answers Brief and Concise. That means no more than a few minutes per response. That is especially possible when you have practiced the likely questions about your background as it relates to the employer's job beforehand. What do you study for practice? Your resume. Prepare to briefly explain skills and experience from your resume as it relates to a description of the employer's job opening. Most people don't do that. You will, and that's why you'll win.

Quantify When Possible. Evidence sells. That evidence may be money that you saved or gained for a previous employer. It can be a statistic related to your success as a student club leader or whatever activity proves you to be a value-added contributor. Whether in a leadership role or serving as an individual contributor, know that employers will be impressed with concrete evidence of your accomplishments and motivation. Give the interviewer evidence, and do it with enthusiasm.

Describe Your Strengths. It is common management wisdom that people should be hired based on their strengths, rather than lack of weaknesses. So, know your strengths and especially how they relate to the employer's needs. When asked, give brief and concrete examples of your strengths in action. Show how your strengths produced positive results. Again, be specific and quantify when possible. You are talking with technical and business people who love numbers expressed in dollars or percentages as evidence of value-added contributions. Remember that in addition to quantifiable technical strengths, employers are looking for the soft skills—your ability to communicate, to work on a team, and to be a potential leader. Together, this represents your package of strengths that you will be communicating throughout the interview.

Prepare for Questions. The interviewer may ask behavioral interview questions. Prepare to be asked questions, as well as to ask questions. This will be covered in next sections of this chapter.

At the End of the Interview. Shake hands with the interviewer, and close the sale. Remember, you are a salesperson and you are selling yourself to the employer. Sincerely thank them for the opportunity to interview. If you are interested in the company and position, be sure to tell them of your enthusiasm and confidence that you can do their job very well. If you haven't yet done so, give them your business card and ask for theirs. Establish any follow-up action that can help them or you come to a favorable conclusion. Let them know that you would welcome an opportunity to contribute to their team. More on closing at the end of this chapter and in Chapter 9.

BEHAVIORAL INTERVIEWING

Behavioral interviewing is likely to be part of the interview. It's based on the premise that your past behavior is the best prediction of future performance. There is an increasing trend for companies to use behavioral interviewing. Therefore, you need to know what this is and be ready to respond to behavioral interviewing questions.

- ❖ Behavioral questions ask you to respond as to how you handled a specific situation that you experienced in the past.
- ❖ Behavioral questions differ from traditional interview questions. The traditional interview question might be how you would handle a situation. Behavioral questions ask how you *did* perform.
- ❖ Here are some examples of behavioral interviewing questions:
  - Give me an example of an important goal that you set sometime in the past and how you achieved it.
  - Tell me about a time when you had a personality conflict with a coworker or student and explain how you handled it.
  - Describe a time when you acted as a leader to complete a team project.
  - Tell me about a major problem you have faced and how you dealt with it.
- ❖ How to be a STAR in a behavioral interview. That's right—you will be a star when you're prepared with a strategy. Remember the word STAR and what goes with it:
  - ***S—Situation.*** Briefly describe a situation most relevant to the question.
  - ***T—Task.*** Briefly, but specifically, describe the challenge you faced.
  - ***A—Action.*** Briefly describe the specific steps you took to solve the problem.
  - ***R—Result.*** The result is always positive, even if just a learning experience.
- ❖ Preparation for the behavioral interview; prepare to be a STAR.
  - Review your resume and recall situations involving favorable or unfavorable situations likely to be used as examples in a behavioral question. Behavioral questions of engineering students commonly involve coursework, work experience, teamwork, leadership, initiative, and planning.
  - Write a brief summary of each situation, using the STAR model.
  - The positive action you took and especially the result are the keys to a successful behavioral interview question response. Again, let me

emphasize that your result must *always* be positive. Even if you were asked to describe a situation in which you failed—acknowledge the failure, but the result was that you learned something valuable. Specify what that valuable thing was and how learning it has served you well ever since.

TRADITIONAL INTERVIEW QUESTIONS

Prepare yourself for brief answers to these common questions and you'll be way ahead of your competition. Answer these questions concisely and, if possible, relate your answer positively to the employer's need.

Personal

Tell Me About Yourself. This could get you tongue-tied if you're not prepared. But you love this question because you're ready for it and others are not. The response is not "Whaddaya wanna know?" For example, instead say "I was born and brought up in (location). I discovered that I was good at and enjoyed math and (chemistry) in high school, and therefore decided to study (chemical engineering) at ________ University. I've (enjoyed/done well in) my coursework and labs. I also believe that I have leadership qualities. I was (officer) in the student chapter of (AIChE). My favorite subjects of (name them) relate to the technical needs described in your (position) opening. I feel qualified for (the description of your technical requirements) and like your emphasis on teamwork."

Why Did You Choose to Interview with Our Organization? You've done your homework; you know what the company does and what the job requirements are. This is your opportunity to tell the interviewer why your skills and interests match their requirements. Let the interviewer know that you have studied their organization and really like what they're doing. Show your enthusiasm and knowledge. Let the interviewer know what you can offer them.

What Are Your Greatest Strengths? You are prepared for this question. You will be able to confidently summarize your strengths shown on your resume without bragging. Do so by stating facts of accomplishment and how those strengths allowed you to succeed. It is important to know and explain strengths that are most needed by this organization. Have several key strengths in mind, along with examples of results achieved through your strengths. For engineering employment, it is generally necessary to have technical strengths along with soft skills in the functional area for which you are being considered.

Describe Your Weaknesses. Prepare to be asked for any of your weaknesses. The fact is everyone has weaknesses, and you have to *carefully* select ones that you will describe to an employer. The weakness you describe should be a form of a strength. A reasonable weakness for an engineering student is that you may lack industrial experience. If so, describe your relevant coursework and emphasize that you are very motivated to get additional experience available with this employer. You could describe a weakness as "I'm somewhat impatient." That can be interpreted as a strength that you're motivated and anxious to get things done. If you are in fact impatient, you want to describe how you control that behavior so as to not alienate team members. Similarly, you might say "I'm a perfectionist." That says you insist on high-quality work, but it's also a red flag that you are a "paralysis of analysis" victim. Better in fact, to say that "I've had perfectionist tendencies in the past but learned to move ahead on projects and to overcome unnecessary delays"—using an example to prove that point. Another common challenge for students is time management. You could acknowledge that earlier in your college experience you had trouble with time management, but since then you have become well organized, using a time management tool. This could be an electronic or paper calendar/organizer that you might quickly show the interviewer. Public speaking is something nearly everyone without experience fears. You could admit that weakness, especially if you are overcoming the fear by being enrolled in Toastmasters, a speech course, or similar activity that shows your determination to overcome any obstacle.

Describe One or More Accomplishments You Are Really Proud Of. As an engineering student, you can be justifiably proud of your acceptance and success in an engineering school. Be specific. You could describe your academic accomplishments or your leadership or teamwork accomplishments. Employers are looking for people with success in both technical and soft skill areas.

Employers also seek creative, result-oriented innovators. If you have done anything new, different, and valuable, be sure to describe it along with any quantitative value produced.

Education

Why Did You Choose Your Major? This is a good opportunity to explain your strengths and interests that led you to your engineering specialty. Impressive candidates express enthusiasm about their field of study. Moreover, if you can relate your major to the needs of the company interviewing you, that is a special bonus.

Which Subjects Did You Like Best? Your best subjects are quite likely to be within your major. Describe the subjects that you did well in and how they might apply to the job opening. During an employer's location interview, this question might lead to more in-depth questions about your knowledge and application of your favorite subjects.

Do Your Grades Accurately Reflect Your Ability? Obviously, if you have a 4.0/4.0 average, you'll say yes; if you have a 2.0/4.0, you'll say no. But explain further: if you have a 4.0 average, demonstrate that you have soft skills beyond your technical excellence. If you have a 2.0 average, especially emphasize your other skills, perhaps hands-on achievement rather than theoretical. Leadership and other people skills might be your strength, rather than technical coursework. Wherever you are on the academic spectrum of grades, sell your strengths and convince your interviewer that you are applying for a position that suits your capabilities.

Did You Participate in Any Campus Activities? Employers seek well-balanced engineers who have done some things besides studying. Teamwork and leadership are key assets. If you were in an engineering club or sports team, describe any leadership or teamwork results you've achieved.

How Did You Pay for Your Education? If you had scholarships or worked part time, be prepared to explain this. Don't be embarrassed if your parents paid your way—but it's good to have an answer that you have taken the initiative to help shoulder the expenses of your education. A co-op assignment is ideal to demonstrate the industrial experience you've gained while helping to pay for your education, so share your specific experience.

Experience

What Job-Related Skills Have You Developed? Relevant skills could have been gained from employment or a school laboratory, club, or activity. Describing your co-op experience is ideal, including any quantifiable results. Use the STAR model described previously to summarize your job objectives, action taken, and results achieved. Employers are especially interested in your:

- Technical skills—including application of theory, hands-on experience
- People skills—those acquired through customer service or leadership
- Initiative—where you are able to demonstrate your self-starting ability

Tell Me About Any Jobs That You Had While in School. What Did You Learn From These Experiences? Be prepared to summarize your experiences and results. Tell what you learned, as well as the value you have contributed for your employer. Again, if possible, relate your skills to the job for which you are applying.

Give Me an Example of a Situation Where You Provided a Solution for an Employer. You are prepared for this behavioral question. Your preparation has included being able to give some specific, positive results on items listed under "Experience" in your resume. This gives you another opportunity to use the STAR technique and be a star in your response. Being able to quantify a positive result would be especially good. The interviewer is looking for evidence of your ability to solve a problem—whether it's a positive result for the customer, money saved, or dollars generated.

How Do You Think a Former Supervisor Would Describe Your Work? This question could throw someone who's not prepared. "Good" is not a sufficient response. The best answer would be giving an actual example of an occasion where you performed well and received a compliment for your work. In fact, that would be an opportunity for you to refer to your list of references and suggest that a former supervisor be contacted to confirm your positive qualities and results achieved. Remember, at all times you are selling yourself and you want to take advantage of every opportunity to prove your value to the interviewer.

Career Goals

What Are Your Career Goals? This is big. It's a big opportunity to rise or fall in the interview. And it's an important question often asked. Most people haven't given their career goals much specific thought, so your intelligent response will move you well ahead. Study your self-assessment in Chapter 3; it will enable you to give a succinct summary of your tentative career plan, for example:

- *Short Term:* Establish my technical foundation and performance track record in (your area of peak capability and passion).
- *Long Term:* Explain your reasons for choosing your intended path of career progression. It could be the technical specialist ladder or management, if appropriate based on your self-assessment. You choose your targets, in line with your ability and motivation.

The key in responding to this question is that you give a thoughtful answer that indicates that you know how to plan a project—your life. It's also best if you express your career goals in light of what the company interviewing you can offer.

Where Do You Want to Be 10 Years From Now? (20 Years?) This is a variation of the last question. Your response would be the long-term goal answer to the previous question. Again, think of your answer alongside the opportunities available with this potential employer. Interviewers are thinking of your future plans alongside their company's future needs and hoping there might be a fit in the long term. Here are some answers that *won't* impress:

- I don't know.
- I want to be a stay-at-home mom/dad.
- I want to go to New York City and be a dancer.

Those could be honest answers, but they won't help you get hired. I'm not suggesting that you be dishonest; just be aware of potentially self-destructive responses.

How Do You Think Our Company Would Contribute to Your Career Goals? Refer to the previous two questions. Your response should focus on your skills, interests, and values relative to the employer who is interviewing you. Let them know, specifically, how it would be a win-win both in the short and long term.

Personal Preference Questions

Prepare for questions about travel, relocating, and working overtime. All three of these are common requirements with many companies. Don't apply for jobs that don't fit your personal requirements. Apply for appropriate jobs, based on your individual goals and situation. Note that you may need to adjust your lifestyle to get the job you want. For example:

How Do You Feel About Working Overtime? Professional jobs increasingly entail more than 40 hours per week. Therefore, the company might like you to answer that you're committed to getting the job done and can work overtime as necessary. You don't want to be seen as a clock watcher who always wants to bolt out the door at 5 p.m. However, neither do you want to get sucked into an unexpected situation where you accept an eight hour a day job and end up regularly working 12+ hours per day. It's important for you, as an engineer, to be flexible and committed to the job. Meanwhile, you should have a clear understanding of what job commitment means to the company alongside your personal needs.

Are You Willing to Travel? Are You Willing to Relocate? As with the last question, it is appropriate to ask for more details as to frequency and length of the travel requirement. Also, know about the frequency of relocation requests. Neither you nor the company can have an employment relationship where you are unable to fulfill the job description. Be aware that if you aspire to be on the management fast track with a large corporation, it is common to get broad experience on big assignments that involve travel, relocation, as well as overtime.

How Do You Feel About Working on Several Assignments at Once? Give Me an Example of How You Have Done This Successfully. Yes, you can only do one thing at a time at the moment. But often there are multiple priorities and projects. If you are good at multitasking, say so and use the STAR model to prove it. Some people love to multitask, others don't. You owe it to yourself to seek an environment that suits your ability and temperament. Regarding this question, some degree of multitasking is required in most engineering assignments.

Would You Be Successful Working on a Team? Why? Give Me an Example. As a general rule, employers want engineers who can work on teams. You will see in later chapters, it is increasingly common for engineers to work on project teams. Therefore, any athletic, club, or other team experience you have had is likely to receive positive consideration by an employer. Tell them about your team experience and give them a positive example of how you contributed.

What Kind of Boss Do You Prefer? In answering this question, understand that more and more bosses are trusting and empowering their employees to work without close supervision. Again, there is increased emphasis on teamwork with team members empowered to do their part of the project. So, if you have high initiative and can self-manage your time and projects, that is a big plus. If high initiative describes you, then your answer might be "an empowering boss who is committed to the professional development of the team."

What Is Your Salary Expectation? You may not be asked this question, but then again, it could pop up, and you will be prepared. Actually, the longer you can put off answering with a specific dollar figure, the better; for it is best to focus on your qualifications and the needs of the company so that you can establish your value to that organization and get them to want to hire you before discussing salary. If you answer early with a dollar figure that is too low, they might take advantage of you; if your answer is too high, they might feel unable to afford you. If pressed, you can let them know from your research that a (your major) in (location) ranges from ________ to ________ based on qualifications, and that you trust that they would be competitive in a salary offer based on your academic performance and experience. Then, hopefully the conversation will get back to how you can be the best candidate to meet their needs. Ultimately, the company is deciding whether you are a good business investment who would solve problems, create opportunities, and contribute to their bottom line financial results. If they like you, they'll pay you as much as they can.

ARE YOU ABLE TO THINK ON YOUR FEET QUESTIONS

Job Related, Technical

When being interviewed at a company's location, it is not unusual for interviewers to ask a candidate how to solve a problem that relates to the organization's business. They are trying to see whether you have the technical skills and aptitude to hit the ground running on one of their assignments or whether you would require a lot of training from them.

Preparation. Review the company's products and the technology involved in producing these products. If you are an electrical engineering student, you may be asked to go to the white board and diagram an electronic circuit or power system. A variation of demonstration questions could be asked for any major. The interviewer needs to know your level of ability to apply your theoretical coursework. The employer wants to know what you know and especially what you can do with what you know, along with how you can communicate and think on your feet.

Specific Engineering Functions

Preparation. Your ability to think on your feet will be the key to this one, whether its design, sales, project engineering, manufacturing, customer service, and so on. There are Internet sites and books to give you an advance look at what these functions include and the qualities employers seek. Conversations with specialized professionals in your network will also be very helpful in preparing for functional questions.

For example, a manager interviewing a potential sales engineer may hand you a pen and say "sell me this pen." Don't panic—think. There's no one answer to this request; customer focus is very important in all engineering fields, but directly in sales, including the ability to understand the customer's need. Therefore, a good answer to the pen question might begin by asking the "customer" what is required in a good writing instrument and the environment in which it will be used. Is the focus on high-quality or low-cost? Fancy-office or dirty-construction site use? Then you might demonstrate how the pen would or would not meet the customer's need including suggested hypothetical options, if appropriate. Finally, you let the customer know that you have a supply of thousands of these pens to provide for the whole company. Invite the interviewer to place an order for the pens. You've just gotten an A in a behavioral demonstration question.

Here's another challenging scenario for a sales engineer candidate: The ability to persevere, handle stress, and rebound from rejection are important qualifications for a salesperson. So, for example, you as a sales candidate might interview well throughout the day—in fact so well that you get to meet the big boss at the end of the day. Great! You're heading into the big corner office expecting to receive a job offer. But surprise! The big boss is standing beside the desk looking at your resume. Instead of warmly shaking your hand and inviting you to sit down, the boss just looks up at you and says "you're not qualified for this position." What do you do? Apologize and leave the room in a hurry? Start swearing? Burst into tears? You are in the big boss's office only because you have done very well in all of the interviews throughout the day—and the boss knows it. Actually, you're facing the final step before being given a job offer. This is to test your ability to handle a stressful situation—stress and rejection. The successful expectation is that you will come up with a response that respectfully, but firmly, points out that you are qualified for the job for the following reasons. Add that you are highly motivated to work for this company and give reasons that you can create value. It's unlikely that as a new graduate or co-op student you will have an interview experience that is this stressful, but I include this as an example so you can prepare for the worst and come out on top with a job offer. What the big boss was really asking is a combination of three questions: Why should I hire you? How do you handle rejection? How do you handle an actual unexpected stressful situation? If you are not prepared for almost anything, you could be thrown off balance and lose out on a great offer. Often, the best opportunities are preceded by a challenging screening process. This example was for a sales function, but relevant interview-demonstration competence testing should be anticipated in any function for which you are interviewing.

Off-the-Wall Questions

These questions are unrelated to the job or your coursework that you cannot study for, except to know they may be coming, to prepare to be logical, and to not panic. They test your ability to handle unexpected, crazy questions and figure things out in a logical, practical way.

Why Are Manhole Covers Round? Your answer should not be what does this have to do with the job I'm applying for? Answer the question using your engineering, logical thinking. There could be several answers:

- Because a circle is the only shape that can't fall into the manhole.
- Because the manhole is round.
- Because manhole covers are heavy and a worker can roll them.

Giving more than one answer could score extra points for your logical reasoning in response to a question you've never even thought about before.

How Many Gas Stations Do You Think There Are in the United States? Wow! How the heck would anyone know? The interviewer doesn't want to hear that anymore than if you were to say it's a stupid question. The interviewer is seeing how you might logically dissect a problem without having hard data, using a sequence of assumptions.

- The population of the United States is a little over 300 million.
- Let's say there is a market need for a gas station for every 1,000 people.
- Based on that assumption, there are approximately 300,000 gas stations in the United States.

It's a problem-solving estimate that you reason out step by step on the spot. The important thing to the interviewer is it you take a logical series of assumptions, breaking the problem down into components. The unprepared candidate would give up or make a wild guess, without revealing a logical thinking process.

How Many Piano Tuners Do You Think There Are in a Metropolitan Area of One Million People? He doesn't know and he knows you don't know. What the interviewer is looking for is the same logic flow.

- There are one million people in the metropolitan area.
- Let's say there is an average of four people per household; that's 250,000 homes.
- Let's say there is a piano in one out of ten households, that's 25,000 pianos.
- Let's say that a piano is tuned every other year, that's 12,500 pianos to be tuned each year.
- Let's say a piano tuner can tune an average of four pianos in a day.

The interviewer will probably move on to the next question after your second or third assumption. The precise answer is not important; you've got what the interviewer wants—the ability to tackle and size up a problem by breaking it down into logical, sequential steps. Obviously, on an actual engineering project you would use precise data, not guesstimates, but this ability to have a logical flow of thinking is very important.

You, a Boat, and a Rock. Here's one that calls for your ability to apply elementary physics in a hypothetical situation. You are sitting on a pond in a boat with a large rock in it. You manage to throw the rock overboard and into the water. Presuming you could measure it, does the water level in the pond go up, down, or remain the same? Why? Think before you answer. Don't rush.

- When the rock is in the boat with you, it displaces its weight, causing the boat to sink lower in the water than if it was just you in the boat.
- After you throw the rock into the water, it displaces its volume.
- Therefore, the water level of the pond would go down, since the boat would have risen, displacing less water.

Separate a Truckload of Rocks and Wood ASAP. Choose any kind of truck and any kind of circumstance.

- You've chosen a dump truck for your mixture of rocks and wood and chose to put your dump truck on the boat landing of a lake.
- You will cleverly do an immediate separation of the rocks and wood by dumping the load into the water—you won't have to remind the interviewer that wood floats.

The specific questions in this section are not likely to be asked, but it is wise to prepare for the range and flavor of these off-the-wall questions. They are relevant to how good engineers can think on their feet and use common sense and logical thinking, rather than giving up, in an on-the-spot situation. The rock in the boat question has been an actual favorite of an engineering manager to see if the candidate was able to recall and apply elementary physics.

THE INTERVIEWER'S TURN TO TELL AND SELL

Let's presume that you have impressed the interviewer with your responses to her questions. So far you've sold yourself to the interviewer. Based on the favorable impression you have made, the interviewer might exercise the option to do some selling. Specifically, she may give you more information about the company and the specific position. The purpose? The interviewer is sold so far and wants to give you positive information so you might seriously consider a possible employment offer. Together, you work to fine-tune whether your skills and interests match the company's needs. Suggested steps at this phase:

- Interviewers will talk and you should intently listen. It is acceptable, even desirable, to take notes on your pad concerning their explanation of the key elements of the company and the job.
- While listening, make note of any job-related questions that the interviewer does not cover during the explanation of the company and the position. You will wait and ask these questions following her presentation of the opportunity.

- The interviewer might give you a continuous sales pitch for several minutes without inviting questions. Another possibility is that she might pause at intervals and ask you for your response, in particular, your interest and qualification for the specific subject at hand.

ANY QUESTIONS? NOW IT'S YOUR TURN AGAIN

Expect at the end of the interview to be asked whether you have any questions. Be prepared for job-related questions not covered in the company literature or the previous interview discussion.

Ask About

- *The Position.* It is appropriate to ask any specific questions about the position that may not have been explained by the interviewer. Obviously, don't ask if it has been covered during the interview, but—if needed—ask for more details of what technical or other qualifications would be required of you so that you can judge whether this position fits your skills and interests, and thereby hiring you would represent a good investment for the company. If there is a match, say so, and with enthusiasm.
- *Skill Requirements.* If this has not been sufficiently covered until now, you want to be sure that you understand what the interviewer views as the most important skills and personal qualities for the position. Then you can explain to the interviewer how you meet these job requirements.
- *Career Development.* It is relevant to ask about career development, including training and advancement within the organization. This can be combined with a question of whether there is a performance evaluation system in place to recognize and develop employee performance.
- *Location and Travel Requirements.* If you have personal restrictions with regard to these subjects, now is the time to discuss them. Significant restrictions could impair your further consideration; however, if you have personal constraints, it is best to see what is available within your limitations. On the other hand, if you are flexible on these issues, make that known—it could be to your benefit. Many large corporations seek people who are willing to gain experience and be promoted to positions involving relocation and travel. International assignments, including Asian, are increasingly common as a step on the career path for high-potential employees.
- *Timetable for Filling the Position.* You can inquire when they plan to make a job offer decision and when a response would be expected of the recipient (potentially you). That will help you place their timing alongside the schedule of other employers you are considering.

- ❖ ***Confirm Your Understanding of Key Conclusions.*** Whether or not the above has been covered, refer to your notes and, in brief summary, confirm your understanding of key conclusions that have been reached during your discussion. Center especially on why you are qualified for the job. This confirmation becomes a question as well as a statement of understanding. It shows the interviewer you have paid attention to key issues about the job, the company, and your qualification. It also permits either of you to clarify any misunderstanding. You have just proven yourself to be an active listener, a good communicator by checking for understanding, and very professional.

Don't Ask About

- ❖ Subjects previously covered.
- ❖ Information available on the Web site or in their annual report—that would show that you did not do your homework.
- ❖ Benefit plans or salary prior to an offer being made. The employer will initiate this information, usually after the evaluation interview. Then, after you receive an offer, you can ask questions and, if necessary, try to negotiate.

AT THE END OF THE INTERVIEW—CLOSE

Once again, remember: you are in sales. During the interview the employer has been evaluating you. Meanwhile, you have been evaluating and selling yourself to the employer. The interviewer is in charge, but from your standpoint, close on a positive note.

Do this by shaking the interviewer's hand and thanking her for the interview and if you're genuinely interested, express your enthusiasm for the organization. State some key reasons that you believe you are qualified for the position. Convey your desire to be a part of their team. Exchange business cards and ask if they mind if you follow up with them in a few weeks.

AT THE END OF THE SITE VISIT

Ensure that the employer has a list of your references and any other materials required for their decision. Presume the employer will check your references before extending an offer. Be sure you know what the next step is before you leave, including any action that is required by you.

THE INTERVIEWER'S EVALUATION OF YOU

Your interviewers will generally complete a form, evaluating you based on their observations during your discussion and judging you on typical qualities. The following is a sample scale that may be used to rate you.

INTERVIEWER'S EVALUATION OF YOU

	Outstanding	Good	Average	Fair	Poor
Professional appearance					
Communication skill					
Technical qualification					
Experience qualification					
Defined goals					
Attitude					
Friendliness					
Enthusiasm					
Problem solving skill					
Teamwork skill					
Leadership potential					
Fit for our organization					
Overall evaluation					

Summary comments ________________________________

____

____

Recommendation _________________________________

____

A form such as this could be used for a campus interview or an employer's location interview. The campus interview would be more brief (~30 minutes); the employer's location interviews would typically allow you to meet with a series of interviewers who would go into more technical depth concerning their opening and your qualification for it. The site interview is designed to allow you and the employer to mutually assess how you would fit the specific position as well as the organization in general.

STUDY THIS CHAPTER BEFORE YOU GO INTO YOUR INTERVIEWS

Why? So you will be at your best during the interview. You will be more prepared, less nervous, less surprised, and certainly more likely to nail down an offer. There is going to be competition for the best jobs; through preparation, you can come out on top. Remember: PPPPP—prior preparation prevents poor performance.

Chapter Summary: Takeoff Tips

What is the employer looking for? Why should they hire you?

- ❖ Competence—your ability to do the job
- ❖ Motivation—your interest and passion to do the job
- ❖ Personal qualities—those that will enable you to fit well in the organization

It is crucial to approach the interview with clear focus on:

- ❖ The employers' specific job requirements
- ❖ How you can fulfill those requirements with confidence and enthusiasm

Interview tips:

- ❖ Know yourself and know the employer.
- ❖ Dress to impress.
- ❖ Be friendly and sincere, with evidence that you can meet the employer's needs.
- ❖ Prepare to respond to this chapter's interview questions, including those in the following exercise section.
- ❖ Prepare for behavioral interview questions, using the STAR technique.
- ❖ Prepare for technical questions related to your major and the job opening.
- ❖ Be prepared to think through an off-the-wall question.
- ❖ Visualize a positive interview.
- ❖ Prepare one or more job-related questions.
- ❖ Close at the end of the interview. Remember, you are in sales.
- ❖ Ensure that the employer has your references and any other required materials.
- ❖ Review the "Interviewer's Evaluation of You" sample form. Be prepared.

Exercises

1. What, generally, do employers seek in a candidate?
2. What steps do you take to prepare for an interview?
3. Answer these important interview questions:
   a. What are your goals?
   b. What are your greatest strengths?
   c. What are your weaknesses?
   d. Why should I hire you?
4. Define behavioral interviewing. Why is it used?

5. How can you be a STAR in a behavioral interview?
6. What are appropriate questions for you to ask at the end of the interview?
7. Tell about a successful interview. What did you learn?
8. Tell about an unsuccessful interview. What did you learn?
9. Are you prepared to be interviewed? What else do you need to do?

Looking Ahead

Many students send resumes out or are interviewed, then fail to follow up. Next, Chapter 9 explains why the follow-up action is a necessary part of your action plan to ensure that you obtain the interviews and job offers you seek. You are taking all the right steps, but the campaign is not over until you have interviews and a job offer from the right employer. Persevere. You have more control over the process than you may think.

CHAPTER 9

Follow-Up Action

WHY IS FOLLOW-UP NEEDED?

After the resume goes out or the interview takes place, you need to follow up. Why?

> **Observation #1.** Many job candidates send a few resumes out, then wait for an invitation to interview—with no follow-up action on the initial resumes to obtain an interview. Further, these candidates tend to not continue to pursue other employment opportunities naively hoping that the first resumes will score—or dejectedly concluding it's hopeless.
>
> **Observation #2.** Many job candidates who are interviewed fail to follow up with thank-you correspondence or any kind of self-generated action following the interview to close the deal on a job offer.
>
> **Observation #3.** Because of their failure to follow up, many candidates will lose out on an interview or a job offer. It doesn't have to be that way!

Whatever your next goal is, an interview or job offer, continued action and discrete persistence are part of the game. Presuming you want to win, you will take the extra steps needed to be sure you are on a positive action track with potential employers.

AFTER SENDING RESUMES

After sending a resume to an employer, generally wait for two weeks before following up; you don't want to be seen as desperate or to be unpleasantly hounding the employer. Consider these alternatives:

- ❖ ***Look for a Network Contact.*** This could be your first approach for application, but for follow-up you may know someone who works for your target employer. Contact this person to see if you can be connected with the line manager who has the opening or with human resources. A network contact who knows you well might be able to make a powerful recommendation to the right people that will result in an interview.

- ❖ ***Another E-Mail.*** You can send another e-mail expressing significant interest in the employer's job posting or however you learned of their position. Include in your message why you are qualified for their opening in somewhat different language than your first cover letter. Attach a copy of your resume again "for their convenience" and express the desire to meet with that employer to discuss how you can create value for their organization.
- ❖ ***Your Career Office.*** Check with them for suggestions. They may know the customs or preferences of the employer for follow-up. If they will be interviewing on-campus, get advice as to how you can best position yourself to be selected for an interview.
- ❖ ***Keep Applying to Other Employers.*** You may have selected some employers that especially interest you. Don't stop there. While you will be following up with initially targeted employers for an interview, explore other possibilities as well. Those organizations that you have chosen at first might not select you for an interview or offer. You need to understand that is a possibility, along with the need to continuously move on with applications to other employers. Don't stop looking and applying until you have accepted a job offer.

AFTER THE INTERVIEW

First of all, nice going! You've been interviewed!! Please feel good about the fact that you have produced a winning resume following your self-analysis and market research. You attracted a potentially right employer to interview you, based on your skills and interests. You've accomplished a lot just to get the interview. Moreover, I'll bet that before the interview, you reviewed your self-analysis alongside the employer's needs, specifically their job opening. You studied the last chapter, knowing the range of questions that you would be asked—with the result that you feel that you did a good job in the interview. Great! Yes, there were a few questions you woulda, shoulda, coulda answered better—but make note of these items for future improvement. Also, don't beat yourself up in that the interview wasn't perfect. Nobody's perfect, and that includes you. In fact, you can get hung up on perfection; "perfect is the enemy of good." Good keeps you moving ahead. Perfect, as noted in the last chapter under weaknesses, can result in stepping on the brakes, leading to paralysis of analysis. So, after your good interview, let's continue to step on the gas toward a job offer.

Written Summary

Summarize, in writing, the interview and action steps ahead. It is helpful to summarize what happened in the interview right afterward, including your impression of the employer and their apparent impression of how you would fit into their organization. Make specific note of what you can contribute to that project. If you exchanged business cards with any of the interviewers, note them by name and any positive conclusions reached. Note any follow-up action required by you or by the employer. Do you and they have all the information needed? When are they going to make a hiring decision? Do you want to work for this organization? How excited are you about them?

Thank-You Notes

Write a thank-you e-mail to each interviewer. You may have only been interviewed by one person, or you might have contact information for only one person. In any case, send a sincere, concise message to each interviewer, within one day of the interview.

Thank the employer for the interview and, if appropriate, express your strong interest in working for the company. You could send your e-mail thank-you to the key contact interviewer, asking that she share your gratitude with other interviewers. If you had specific technical discussions with other interviewers, you could customize a message to more than one of them, tailored to the content of your discussion and especially how you might fit their need.

You can send your thank-you message by e-mail, as it is now the primary mode of written business communication. A nice e-mailed thank-you message can also be forwarded by the key interviewer to other decision-makers in the organization.

Your thank-you message might include something like this:

Dear ____________,

Thank you very much for interviewing me today. Our discussion gave me enthusiasm and an extremely positive feeling about how I could contribute to the needs of (name of employer).

I believe that my (specify skills) and interests in ________would allow me to contribute well in your (technical project area). For these reasons, I am very hopeful to receive an employment offer from you. Given the opportunity, I will work very hard to create value for (company). If you need any further information, please let me know. I look forward to hearing from you.

Sincerely,
Your Name
(987) 123-4567
yourname@internetprovider.com

As indicated, this message gives you the opportunity of commenting on any specific positive interaction you had with each interviewer. Make note of your motivation to work for the employer and your ability to contribute.

Subsequent Follow-Up

A week or more after your interview and thank-you message, consider phoning or sending another e-mail to the key employer contact. Use good sense and sensitivity with subsequent follow-up. You want to express your continued enthusiasm to receive a job offer and make a contribution for the company. Meanwhile, however, you don't want to be pestering the employer or appear to be desperate. Use good judgment, based on your feel for how this employer would be receptive to continued follow-up. Again, don't bug them, but don't let them forget you.

Here's a sample of a second message you can send by e-mail or telephone:

Dear ____________,

Pardon my enthusiasm, but I need to tell you again how excited I am to be a candidate for your (position). I believe that I have the competence, motivation, and teamwork skills to make a strong contribution to your (project or company). Attached for your convenience is another copy of my resume. If you could let me know when you plan to make a decision, I would greatly appreciate it. Thank you again for your consideration.

Sincerely,
Your Name
(987) 123-4567
yourname@internetprovider.com

Closing the Sale for a Job Offer

It seems that most engineering students are not interested in being salespeople. However, the most successful engineers are aware that, regardless of their title, they will be persuading, inspiring, leading, and—let's face it—selling themselves and their ideas throughout their careers. That's how engineers get things done. And that's why, from preparing your resume through the interview and follow-up action, you will keep in mind ABC. What's ABC?

ABC = Always Be Closing. This is a sales fundamental. Closing means that at every reasonable opportunity connect the product or service (you) to the customer's (potential employer's) need. Most engineering students who are disinclined toward sales would not have the instinct to close. But if you connect your strengths and interests to the employer's specific need, you can make a very strong impression of being the most competent and motivated candidate. That impact can lead to a job offer. In fact, it can lead to multiple job offers. Ideally, you might have more than one employer competing for your services. In that case, you'll be following up with multiple employers with the hope that you will receive job offers at about the same time, allowing you to evaluate the offers in parallel and give prompt responses to each employer.

Use Discretion. Some salespeople can over-close. They can get carried away with enthusiasm and perhaps become too aggressive in their sales approach. Too much aggressiveness is a turn-off to the customer. Engineering students use good professional judgment in explaining how your skills and interests can meet the employer's need without going overboard and are not likely to fall into this category.

Evidence Sells. At the end of the day, when the employer is making a hiring decision and comparing candidates, what will be the key? The key will be evidence of your competence to fit the organization that will determine whether or not you will receive an offer.

- *Tangible Evidence.* That is, your evidence of measurable competence including academic accomplishment such as grades or other school activities. Also important are specific results you have achieved in your employment experience. Finally, there is evidence that you have shown during the interview process as to your ability to competently respond to technical job-related questions, as well as soft skill questions.
- *Emotional Evidence.* Beyond your grades and other quantitative measures of your ability, both the employer and you will experience a gut reaction as to whether there is a good fit and whether you belong together. Do you like each other? This is very important for you both. According to your personality, energy level, and style: do you fit the culture of the company? Beyond the technical aspects, do they see you and do you see them as a likely match? It is good for you both to give this serious consideration as you each make your decisions.

Be Proactive in Pursuit of the Close. As much as possible without being too pushy, maintain transaction control from the standpoint of getting the employer to commit to some sort of action rather than leaving you dangling. One example of this is getting the employer to agree that you can re-contact them, rather than being left with "don't call us, we'll call you" where you are helplessly waiting for them to make the next move.

THE JOB OFFER

The employer has selected you as the most qualified candidate to receive a job offer. All your preparation has paid off. Celebrate, especially if this is the employer you choose to work for. Here are some final steps:

- *Get the Offer in Writing.* You may receive an employment offer by telephone, but don't give a firm acceptance before receiving an offer letter with the salary and benefits spelled out in detail.

- ❖ ***Understand the Benefit Plans.*** You will probably receive standard information from a human resources representative describing the employer's benefit plans; including vacation eligibility, health insurance, possible group life insurance, and any pension, retirement savings, or other financial investment plan. Be sure to read and understand this information and ask questions about anything that is not clear. Your benefit plans are a significant supplement to your salary, and you should consider them as part of your offer package.
- ❖ ***Relocation Assistance.*** If you are coming from out of town, some employers will assist you with relocation expense. Find out about any assistance that is available.
- ❖ ***Response Time.*** The employer has chosen you above the other candidates and hopes to get your acceptance of the offer as soon as possible. A reasonable employer will allow you one to three days to think about it so that you can be sure of your decision. If you are considering more than one company and prefer one employer over the other, make it a high priority to arrange the timing for an offer from the preferred company to fit within the time limit of any other company's offer.

ADDITIONAL POSSIBLE JOB OFFERS

Ideally, your job search strategy might have it that your interviews would take place around the same time, resulting in possible offers also arriving at about the same time. Don't count on multiple offers; it would be an ideal situation, and if more than one arrives, that's great. If you have two offers from companies that you have chosen as being the right fit for your abilities and interests, now you have the delicious dilemma of choosing between the two (or more, but that is even more unlikely). It may be a no-brainer as to which employer is your first choice. On the other hand, it might be useful to have a logical strategy to evaluate which offer is best for you. Here is an approach that might help you to make a good decision:

- ❖ List the job elements that you need to have—your firm requirements. For example, for a co-op job or new graduate, a job that gives you the best foundation experience for your field. You could also have a financial or geographic requirement.
- ❖ List the job elements that you prefer, but are not essential. List them in order of importance to you.
- ❖ Measure your two or more offers against these listings. Obviously, the elements that you need to have are crucial and outweigh your preferred items. If it's difficult for you to decide between two or three offers, having a listing such as this could be helpful. You can analytically assess the comparative benefits and opportunities of however many employers are competing for your service.
- ❖ Trust your heart as well as your head. You have done a careful analysis of the logical factors associated with the job offers. You've used your brain. Now, check your heart and your gut. Do you want to partner with this

employer? If the logical factors say yes and your heart and gut agree, then you're set. You are prepared to accept the offer with enthusiasm.

A Deal's a Deal

Once you have accepted an offer, you must not continue shopping for other possibly better offers. Your values of honesty and integrity are absolutely fundamental. Occasionally, students will accept an offer because it's the only one available, then try to back out of it when they find an employer that is more desirable. That is unacceptable. Think of it this way: how would you feel if you showed up for work on your first day and your employer told you they didn't need you since they found someone better? That doesn't happen. Neither should it happen that you accept an offer and then decline it later in favor of a better offer.

AFTER YOU ACCEPT THE OFFER

Your Confirmation of Offer Letter. After receiving the offer letter, it is important to acknowledge your acceptance of the offer, restating the details that were given to you in the offer letter. Offer letter confirmation details would include salary, title, location, start date, and so on. Sending this communication will ensure there are no misunderstandings about what you have been offered and provides you a record of an important business transaction.

Pre-Employment Medical Exam, Including Drug Screening. Before you start to work, most employers will arrange for you to have a pre-employment medical examination. If so, your offer letter will state that the employment offer is contingent on you passing the medical examination. This exam will include drug screening to ensure that you are not taking any inappropriate drugs.

CONCLUSION

Congratulations! Your diligence in adhering to a professional, step-by-step job-search has paid off. You have found the right employer and are about to work in a function that aligns with your abilities, interests, and values. A more haphazard approach would not have allowed you to be in the successful career launch position you are in right now.

Chapter Summary: Takeoff Tips

After the resume goes out or the interview takes place, you need to follow up.

- Many candidates send resumes out, then wait for interviews, with no follow-up.
- Many who are interviewed fail to follow up with thank-you correspondence.
- As a result, many candidates will lose out on an interview or a job offer.

After sending resumes, to seek an interview:

- ❖ Look for a network contact, someone who works for your target employer.
- ❖ See if your network contact can recommend you to human resources or engineering management.
- ❖ Keep applying to other employers.
- ❖ Send another e-mail, with resume attached, emphasizing your interests, qualifications, and desire to meet them and discuss how you can add value.

After the interview:

- ❖ Record, in writing, the interview results and action steps ahead.
- ❖ Write a thank-you e-mail to each interviewer.

Closing of the sale for a job offer:

- ❖ ABC = always be closing.
- ❖ At each opportunity, connect your qualifications with the employer's need.
- ❖ But use discretion—some people can over-close.
- ❖ Evidence sells—evidence of your qualifications leads to a job offer.

The job offer—the employer has selected you to receive an offer:

- ❖ Get the offer in writing.
- ❖ Understand the benefit plans and relocation assistance.
- ❖ Determine response time.

Other possible job offers:

- ❖ Ideally, other job offers would arrive at about the same time.
- ❖ To decide, list job elements that you need along with those you prefer.
- ❖ Measure your two or more offers against these listings.
- ❖ A deal's a deal. After accepting an offer, don't continue shopping.

After you accept the offer:

- ❖ Write an offer acceptance confirmation letter.
- ❖ There will be a pre-employment medical exam, including drug screening.

Exercises

1. Give an example of how to follow up in job search:
   - After sending your resume and cover letter.
   - After the interview.
2. To select the right job offer, list (per page 98)
   - Job elements that you need to have.
   - Job elements that you prefer, but are not essential.
3. Explain how you will close the sale for a job offer.

Looking Ahead

Enjoy the fact that you are ready to take off with a good job. The next chapter discusses how you start employment. It will provide information needed so you can have a great start with your employer of choice.

CHAPTER 10

Starting Employment

You got the job! Be proud of your accomplishment. You may be a new graduate, an intern, or a co-op student. Whatever your status, you've succeeded in landing a job with an employer who is hopefully well suited to your skills, interests, and professional objectives.

Remember, your employer has hired you based on evidence that you're well suited to creating value for their organization. Your resume and interview demonstrated that you're competent, motivated, and have the personal qualities that will fit well into their organization. Now you're ready to start work and prove that they were correct in their assessment that you're a winner with a lot of potential to create value.

The suggestions in this chapter for starting employment are made at the high end; that is, some career launch items that the average person starting work might not think about, but are very important for those who want to be the most successful. The suggestions that follow are based on the realities that:

- First impressions are huge; making a good first impression gives you a major head start toward success. The opposite is also true of a poor first impression.
- Inertia (a body in motion tends to remain in motion) also applies to you and the initial impression that you make with your employer. Once you make an impressive, energetic start, both your and your employer's expectations will help you to keep moving ahead successfully. You know the answer to "what happens to a body at rest"—especially if it's on-the-job rest.
- Your habits, good or bad, follow your first impressions and inertia. Pay special attention to your work habits and your personal habits right from the start. It's easy to form good or bad work habits—and they are likely to continue to your career's benefit or detriment.

BEFORE YOU START WORK—PREPARE

- ❖ *Review Your Information about the Employer and the Job.* Review the research that you did about the employer before your interview. That information includes your Web site search about the company and its products, along with your area of work and a description of the job that you'll be filling. Look over your notes following the interview and your employment offer letter.
- ❖ *Contact Your Boss.* See if your supervisor wants to refer you to material pertaining to the job that you could study before coming to work. The boss may or may not suggest that you do any reading before coming to work, but you've impressed your boss that you're anxious to swing into action and show results as soon as possible.
- ❖ *Arrange Your Housing, Transportation, and So On.* If possible, make your arrangements before getting on the job. Do that by agreeing to a start-to-work date with enough prior time to get settled. Pre-arrangements are not always possible, but try to avoid personal getting settled disruptions (i.e., taking time off during working hours on your first assignment). If you're coming from out of town, human resources might be able to lead you in the right direction for housing and transportation information assistance.
- ❖ *Plan How You'll Get to Work.* Time how long it will take; do a dry run timed trip before your first day, and do it during the hour that you would be coming to work. It is not unreasonable to plan one or more alternate routes in case of a traffic tie up or some other unexpected event.
- ❖ *What Will You Wear?* If you visited the employer's location during interviews, you're able to tell how other engineering employees were dressed. In general, dress professionally and conservatively to start. The person offering you the job can give you guidance about any specific attire that is appropriate.

YOUR FIRST DAY—MAKE A GOOD FIRST IMPRESSION

- ❖ *Arrive on Time.* Punctuality in business is imperative. Therefore, a good axiom to follow is "early is on time; on time is late." Arriving to work a few minutes early will allow you to be at your desk, with your computer on and ready for action when the work day starts. Punctuality, or the lack of it, will tend to be a life time habit, and it will serve you well to establish the reputation that you're punctual in arriving at work, meetings, and project deadlines.
- ❖ *Be Appropriately Dressed.*
- ❖ *Be Friendly.* Be naturally cordial to everyone, smiling, making eye contact, and saying "hi" as appropriate. That includes cordiality to a security guard, the receptionist, and people at all levels in the organization. It sends the message that you're a friendly, approachable person and ready to be a good team member.

- ***Learn and Use People's Names.*** You'll be meeting a lot of people during your first days on the job. Name-recall may be a challenge, but the sooner you're able to remember and use people's names when you're dealing with them, the better. Addressing people by their name is a big part of making a positive first impression and getting people to like and respect you. To remember people's names, use their name as you're introduced. Then, as soon as possible, include them on a list you'll make of people you've met and their job functions. Study your list. Make a habit to call the people you've been introduced to by name the next time you see them; that will help you remember their names. People tend to remember things that are important. This is important. Don't say, "I have a terrible memory for names"—that programs your brain accordingly. Instead, say to yourself, "It is important to remember and use people's names, and I'm going to do it." You will.
- ***Have a Great Attitude.*** Show that you're glad to be part of the team and ready to fit in. Show respect to the people you meet and share their enthusiasm about the goals of the organization.

FORMS, PROCEDURES, AND BENEFITS

You'll likely meet with a member of the Human Resources Department to take care of the following administrative details of your employment:

- ***Orientation, Including Employee Handbook.*** The orientation and possible employee handbook will give you an overview of your employer's organization. Following the orientation and handbook guidelines, you'll know the rules, rights, and responsibilities you'll live by while being employed there.
- ***Employee Agreement Form.*** As an employee of a technical organization, you'll probably be asked to sign a statement agreeing that anything you invent or develop while being employed by this organization is the property of the company. You'll also be asked to agree not to divulge any business-confidential information. This is normal and reasonable; you're not likely to have reason to hesitate to sign this agreement. The only exception is if you plan in the future to become an entrepreneur or join another organization to compete with this employer's technology—in that case, revise your employment plans and/or consult a lawyer to avoid getting into a legal snarl with a non-competitive agreement.
- ***W-4 Form.*** You need to complete this form so your employer can withhold the correct Federal income tax from your pay.
- ***Employee Benefits.*** Employee benefits are constantly changing and vary from employer to employer. They are also different depending on whether you're a graduating engineer or an intern/co-op employee. Here are some benefits that are important to know about:
  - ***Health Insurance.*** Find out whether your employer offers you health insurance benefits and what they are. If you're a student on an

internship or co-op assignment, your school's or parents' policies might cover you—but be sure to determine that you have health insurance. As a new graduate, it is especially important to know what your employer's health insurance coverage is, since you need health insurance.

- ***Life Insurance.*** Some companies offer group life insurance to permanent employees at a reduced premium—they pay for part; you pay for part. If offered, it's probably a good deal, but you can check it out with other insurance companies' offerings. At a young age, life insurance is inexpensive; common thinking is you should have it: it's generally offered to new employees without a prerequisite medical exam.
- ***Tuition Reimbursement.*** If your employer offers all or partial tuition reimbursement for graduate school, that's a big benefit. Find out the details of what's offered in this category and consider your career plans alongside the advantage of working while your employer pays your tuition for going to graduate school part-time. Perhaps an engineering graduate degree or an MBA is part of your career plan. If so, investigate.
- ***Retirement Plan.*** Employer-funded pension plans for retirees are becoming a thing of the past. Nonetheless, some employers still have them. If your employer has a pension plan for permanent employees, examine it and find out how long it will take you to become vested—that is, eligible for a reduced pension if you leave before normal retirement age.
- ***Tax-Deferred Retirement Plans.*** Private industry offers 401(k) plans. Public education and some non-profit employers offer 403(b) plans. In either case, these are opportunities to regularly withhold some of your pay to invest for retirement and defer current income tax on that amount. Often your employer will also contribute (supplement) a certain percentage of your pay on a regular basis to your tax-deferred account as a retirement benefit. These tax-deferred retirement plans have become a replacement for the traditional employer-paid pensions. It is important that you take advantage of these plans, along with consideration of a personal Individual Retirement Account (IRA), and especially a Roth IRA. A young person should especially investigate a personal Roth IRA since there is currently no capital gains tax on your investment when you retire. Get educated through books and perhaps a financial planner on the miracle of compound interest and the advantage of starting to invest right now while you're young. Decide on a financial planning strategy that meets your life objectives and your personal investment goals, balancing your tolerance for risk vs. reward. Some people do not fully participate in their employers' 401(k)-private or 403(b)-public retirement savings plan, even when the employer supplements each deposit with a percentage of matching contribution; this could be a big mistake in maximizing the opportunity to build financial wealth over the years.

- *Relocation Assistance.* As mentioned in Chapter 9, if you're coming from out of town, some employers will assist you with relocation expense. Find out about any assistance that is available.

YOUR BOSS AND YOU—SETTING YOUR OBJECTIVES

- You've read the general job description, but now you'll learn the specifics of what your boss expects from you.
- Listen and take notes. Your boss will describe the project being assigned to you and explain the objectives and expectations. This discussion should include the results that are expected of you and when they are expected.
- Your boss will hopefully inform you about the people and other resources available to help you to accomplish your objectives. It would be good if your boss decides to take you around and introduce you to other team members and support functions. Again, take notes, including names of people and functions you'll be doing business with.
- It is important for you to understand what your boss feels is important, and then be able to deliver results. Make your boss's agenda your supportive agenda. Your ability to be promoted is first dependent on your ability to be a good subordinate.
- Establish understanding with your boss regarding the timing and manner in which the two of you will review your progress. While you only have one boss, your boss has many subordinates. Therefore, you don't want to take up too much of your boss's time, and meet with him or her as directed.
- Figure 10–1 is a format that could be useful for you to complete after your meeting with your boss; then show it to him or her for approval of your work assignment objectives and expectations. This allows you and your boss to have mutual understanding about what is expected of you.

Work Assignment Objectives and Expectations

Project Objectives:

-
-
-

Results Expected:

-
-

Action Plan: (including completion dates. Attach additional sheet if necessary)

-
-

Your name ____________________ Date ____________________

Supervisor's name ______________ Date ____________________

FIGURE 10–1 Work Assignment Objectives and Expectations

Do this only if your boss is comfortable with this approach or if this type of assignment definition is part of a co-op program.

Having completed your work assignment and objectives plan, this is a good time to advise you to keep records on every job assignment. Specifically, you could keep the above objectives and expectations plan along with your job title, names, dates, and results for this project as well as all future projects. Throughout your career you want to have an up-to-date resume, and this approach to record-keeping will make the updating of your resume much easier.

YOUR EMPLOYER'S CULTURE—LEARN IT AND GET STARTED EFFECTIVELY

Webster's dictionary defines an organizational culture as "shared attitudes, values, goals, and practices that characterize an organization." You're new. You're expected to eventually bring fresh ideas into the organization, but first you need to fit in. You'll be among a variety of employees, including older workers who have formed enduring habits about how things should be done. Before you spring any of your good ideas about how things should be changed, know how to act. Get off on the right foot.

- Be polite and respectful to everyone, especially older employees with less education who might feel that your youth and education pose a threat to them.
- Listen to the voices of experience and see how things work. Understand how people feel—personally and about the organization. Build trust by showing respect for the knowledge and opinions of others who have worked there for a long time.
- Learn what's going on by reading, watching, and listening. Before suggesting improvements, you need to know the benefit of existing practices.
- Network with people in your immediate area and elsewhere who can help you understand your immediate assignment and how it fits into the big picture. Be prepared to help people in your network as well as to receive help. The name of the game is win-win.
- Be a team player. It's not about you. It's about the team. We rather than I. Get that engrained right away and throughout your career. The less you come across as an ego-driven person, the more successful you'll be. A lot of successful people are ego driven, but the most successful are polite and subtle. Learn the skills of an effective team member and you'll be accepted by your peers and on your way to a leadership position.
- Get a mentor. A mentor is an adviser; a trusted, more senior colleague who can guide you based on his or her experience in the organization. You have a boss; your mentor is your non-boss coach. You might be assigned a mentor, your boss might suggest a mentor, or you can find one yourself. Ideally, your mentor will be able to advise you in various dimensions: technical, organizational, and career. Your mentor, being more senior in the organization, can help you know how to get things done. Having a good mentor in your area of technical and career interest

can be of great help. Your mentor can be a role model who can show you the ropes, introduce you to people who can help you, and be your professional senior friend. Your mentor can connect you with people in their network and enable you to become part of that network. Once again, it's a two-way street: you learn from your mentor and your mentor strengthens their teaching and leadership skills. If you're a new graduate, your mentor can accelerate your career growth. If you're an intern or co-op student, your mentor can be influential in helping you gain a job offer when you graduate.

PROFESSIONAL ETIQUETTE

This is about good manners in a professional environment. It is important that your technical skills are accompanied by personal habits that are more appealing than appalling to your bosses and fellow workers.

- ❖ ***Say Please and Thank You.*** This goes at the top of the etiquette list. The fact is, people are often in a hurry and skip some of the basics even though they know them. Those basics are essential to maintaining the respect and relationships that are crucial in both our professional and personal lives.
- ❖ ***Knock Before Entering.*** That includes whether the door is open or closed. In fact if your boss's door is closed, you won't even want to knock unless it's an emergency or you have prior agreement. This extends to the courtesy of not interrupting a meeting or conversation unless there's an urgent matter that affects the participants.
- ❖ ***Introduce Yourself.*** When you enter an office and meet a person you have not previously met, say "I'm (first name, last name), a (new employee, co-op student, or intern) from _________ Department." Then, state your business. Introducing yourself is a good practice in business, as well as on campus when you're meeting a professor who does not know you.
- ❖ ***Don't Gossip.*** It can be very tempting to pass along a juicy tidbit that will make the rounds of the office, but usually it'll come back to bite you—the person you gossiped about finds out that it was you who spread the negative news. It's not worth it. Gossiping undermines your reputation of professional integrity. If you're known as a person who gossips, people will gossip about your gossiping—that can be unhelpful to your career and it won't add to your list of trusted friends. Yes, you're human, the grapevine exists, and there can be interesting stuff in the grapevine. But try to stay focused on your job. If you incidentally hear things from the grapevine, be a receiver, not a transmitter.
- ❖ ***Avoid Bragging.*** Don't brag about your education, ambitious goals, the important people you know, the expensive car you're driving, big accomplishments, and so on. Bragging is generally not cool. You never know who is going to be offended and sabotage your progress in the organization. Let your work results do the talking, and don't boast of them. Let your boss tell upper management how good you are.

In general, have the poise and manner of a professional. The above items are just a sample of things to do in a professional environment. Watch how your bosses and other successful people in the organization conduct themselves. Your mentor can help.

E-MAIL ETIQUETTE

E-mail professionally, as follows:

- ❖ ***Subject.*** Always have a subject that introduces and titles your message, such as "Request for_________."
- ❖ ***Salutation.*** Address the person you're sending it to as Dear_________ or just the name of your addressee. Find out the e-mail customs of your organization. Don't address your recipient as "Hey _________" or "Yo _________." That's fine for campus or among friends, but not in a professional environment.
- ❖ ***TM or IM Lingo.*** Don't use text messaging or IM abbreviations, such as "lol" or "brb."
- ❖ ***Message Content.*** Use complete sentences starting with a capital letter, and capitalize as appropriate. You as a subject pronoun is "I," not "i." Sending a business e-mail all in lower case is not acceptable. Neither is a message using all capital letters—that's shouting. Begin your message by stating why you're writing and get to the point. Clarify your request of response to any specific action that is required and when.
- ❖ ***Sign Off.*** End your message with, ideally, a pre programmed listing of your name, location, phone number, and e-mail address. In any case, list your first and last name so the recipient will know who sent the message, especially if your e-mail address is not obvious as to who you are. For external e-mails, many professionals sign off with "Best regards," before their signature. More traditional is "Sincerely," then your name on the next line. That formality is generally not used internally.
- ❖ ***Proofread.*** Carefully proofread your message before sending. Check spelling and grammar, including the spelling of the people to whom you're sending your message. Spell-check will help, but your good eyes and judgment are needed to be sure that the message is correct and professional.
- ❖ ***Politeness.*** Professional also means polite, including "please" and "thank you," just as you would do in person as mentioned earlier.
- ❖ ***Copies.*** Send copies of your e-mail message only to appropriate individuals with the need to know; don't bombard people with copies of your e-mail that won't affect them. Also be select in using the "reply all" key. Does everyone really need to know your response?
- ❖ ***When Not to E-Mail.*** E-mail is not the means of settling arguments or sensitive situations. Never blast your frustration or an accusation out in an e-mail. If there is a sensitive issue or a conflict, it's best to personally

meet with the person to resolve the issue. Then, after resolution, send a confirmation e-mail of mutual positive understanding.

PROPER INTERNET USE

- There is a variety of business occasions on which you'll need to view a customer's or supplier's Web site, Google something, use Wikipedia for a quick answer, or use the Internet for other uses directly related to your job. That's fine, especially if it's job related.
- Illegal or immoral use of the Internet can be monitored by the IT department and could get you fired. Know your employer's Internet policies; they are likely to be quite different from your previous environments.

TELEPHONE ETIQUETTE

- ***Answering Your Phone.*** Find out how it is customary for professionals to answer the phone in your organization; then do it as a habit. Rather than just saying "hello," it is common to state your department, then your first and last name. Have a pad and pen nearby to make note of the caller's name and a summary of what the caller is requesting. If you don't know the caller, be sure to make note of the caller's name, organization, and telephone number. Let the caller know what action you'll take and agree on the time required to get back to the person. Then, if you can't handle the caller's question yourself, put it in the hands of someone who can and ensure there is understanding of who will get back to the caller.
- ***Placing Calls.*** First, introduce yourself with your first and last name and, if the person does not know you, what department or company you're with. Be sure to speak slowly and clearly enough that the person you're calling can understand you. Inexperienced persons may be a bit nervous to the extent that they either mumble or talk so quickly that it's unintelligible. Recognize any problems with your communication—if the person you're calling asks you to repeat what you're saying, especially if the caller requests clarification more than once. As when you're receiving calls, make note of any important follow-up action required. And that includes noting who you might ask to take action and noting on your calendar or in your time management system when a response is due.
- ***Voicemail.*** You need to be professional in using it:
  - ***Your Voicemail Message for Incoming Calls.*** Your message can simply be "This is (first name, last name). Please leave a message." Often people have recorded messages that are longer than that, which take up unnecessary time of the caller. Announcing your first and last name is important so that the caller knows that they have reached the correct person, and saying please is polite.

- *Leaving a Message on Voicemail.* It is very frustrating to receive a message from a person who garbles their name incoherently, then gets into a long message, followed finally by "Call me at (a 10-digit phone number)" that is said so quickly that you have to replay the message several times to understand it—if you have the patience to even bother. So here's the suggested improvement on that scenario: say your first and last name clearly, then your organization and phone number, slowly and clearly enough that the person hearing your message can write it down. Then briefly summarize the subject and request the person to call you back. Then state your first and last name again and your phone number—just in case the person didn't catch it the first time. As a student, you might think that is over-the-top silly, but you're ensuring that the person you're calling does not need to keep replaying your message. You'll get a callback and you've left a professional impression.

CONCLUSION

Looking back, you've come a long way since Chapter 1. Now you're on board and ready to start your job. Review this chapter before you begin work. Make a good impression, connect with your boss's objectives, and be aware of the culture and etiquette standards of your organization. You're starting your career on the right foot!

In short, Part 1 of *Ready for Takeoff!* has given you the tools to plan your career and get a job with the right employer based on your skills, passions, values, and goals. Next, in Part 2, we will feature experts from industry and academia who will give you functional tools and suggested opportunities to apply your theoretical engineering knowledge for fast-start, on-the-job career success.

Chapter Summary: Takeoff Tips

Before you start work—prepare:

- ❖ Review your information about the employer and job.
- ❖ Arrange for housing, transportation, and so on.

Your first day—make a good impression:

- ❖ Arrive on time, appropriately dressed and friendly.
- ❖ Learn and use people's names. Have a great attitude.

Forms, procedures, and benefits: take care of the administrative details of your employment.

- ❖ Orientation, including employee handbook. Know the rules.
- ❖ Employee agreement form, including "noncompete" agreement.
- ❖ Enroll in benefits. Don't overlook the 401(k) or 403(b) financial retirement plan.

Your boss and you—set your objectives:

- ❖ Learn the specifics of what your boss expects from you.
- ❖ Make your boss's agenda your supportive agenda.
- ❖ Establish timing and manner in which you and your boss will review your progress.

Learn your employer's culture:

- ❖ Be respectful. Listen, learn, and network.
- ❖ Be a team player. It's not about you; it's about the team.
- ❖ Get a mentor.
- ❖ Use the Internet ethically.

Practice professional etiquette:

- ❖ Face-to-face, interpersonal etiquette
- ❖ E-mail etiquette
- ❖ Telephone etiquette

Exercises

1. Explain how you'll prepare for employment, before starting work.
2. Explain how you'll make a good first impression.
3. Complete a sample work assignment objectives and expectations form, Figure 10–1.
4. Describe how you would learn an employer's culture and become accepted as a team member.
5. Describe the advantages of having a mentor and how you get one.
6. What behaviors of professional etiquette did you learn?

Looking Ahead

You'll now move into Part 2: "Professional Functions and Opportunities." Part 2 begins with Chapter 11: "Overview of Industry: Executive Panel," featuring three chief executive officers and two engineering directors who will tell you their perspectives of what it takes to be successful as an engineer in industry. Part 2 of *Ready for Takeoff!* continues with chapters on project management, value engineering, quality engineering, lean enterprise, engineering professionalism, entrepreneurism, academic careers and graduate school, and becoming a global citizen. Each subject is very important to employers and your career. Each chapter gives you the wisdom of industrial or academic experts and how to professionally apply the concepts.

Professional Functions and Opportunities

PART 2

CHAPTER 11

Overview of Industry: Executive Panel

Big questions: What's it like in industry? What's expected of me? What do I need to know that I haven't learned on campus? How can I be the most successful? Part 2 of *Ready for Takeoff!* will answer these questions by turning to experts in key professional functions and opportunities in the world of engineering employment. We will begin Part 2 with this chapter by going right to the top for answers to some of the big questions posed above. This chapter will give you an overview of industry by featuring five engineering executives, including three chief executive officers (CEOs) and two engineering directors.

These individuals have one thing in common: they have been very successful in starting and/or running an engineering organization. As such, they will give you their executive-level perspectives on what it takes to be a successful engineer or co-op student engineer. Each has a different engineering business and a different story. Since there is no one simple magic formula for success, you will want to seriously contemplate the combined wisdom of these five industrial executives. Then, place it alongside your own skills, values, and action plans as you move toward a successful engineering career.

The five executives will each tell their story and give advice. Then, we will have a panel discussion during which engineering students ask questions of interest and receive a response from the executives.

DOUGLAS P. TAYLOR

Mr. Taylor is President and CEO of Taylor Devices, Inc., a leading manufacturer of vibration- and shock-absorbing devices. He is also President of Tayco Developments Inc. that does research for Taylor Devices. Taylor Inc. shock-absorbing systems have been installed on some of the world's tallest buildings and largest structures. Mr. Taylor holds a BS degree in Mechanical Engineering. He's the inventor or co-inventor of 29 U.S. patents in the fields of energy management, hydraulics, and shock isolation. He's widely published and has lectured worldwide on the subject of shock and vibration. He has received multiple awards for his accomplishments in the field of engineering and is a founding member of the International Association on Structural Control (IASC).

Mr. Taylor offers these perspectives about engineering and you.

Guidelines for Success

- ❖ *Technology Is Changing Fast.* This is one of the first things you'll recognize when you get into business. Ten years after you graduate—whether you're in management, in marketing, or in research, development, or design engineering—you'll have kept pace with continuing advances in technology. If you haven't stayed at the cutting edge in a technical job, you'll be left behind.
- ❖ *Always Be Kind to New Ideas.* It doesn't matter if it's your idea or someone else's. Don't trash new ideas, because you never know—some of the weirdest ideas become the most successful. Example: Mr. Taylor and David Lee were sitting in a restaurant following an Air Force briefing declaring a sharp drop in spending after the fall of the Soviet Union and end of the Cold War. This would have a very negative impact on Taylor's U.S. Defense Department business. So, what to do? Well, let's take the large fluid dampers out of ballistic missile silos and put them in buildings for seismic, wind, and vibration protection. This idea was jotted on a restaurant napkin; 18 years later, more than 300 buildings and bridges use this technology and Taylor Devices has grown by a factor of five.
- ❖ *You're in Business to Make Money and Have Fun.* Most people just pick one, either money or fun. If you're really successful, you can do both. Mr. Taylor makes a lot of money and has a lot of fun. You can too.
- ❖ *Be Aware that You'll Need to Work with Non-Engineers.* As an engineer, your organization will include people who don't directly contribute to your technical work output. They are employed for a reason, and you'll need to work with them. For example, cost accountants may not give you all the money that you'd like for your project. Lawyers may tell you that what you're doing is too risky. Listen to their views and respect their suggestions based on their merit. On the other hand, your project may be obstructed by someone in the organization who doesn't understand the potential of your idea. In that case, you have the responsibility to escalate the issue to a higher level of technical management.
- ❖ *Mind Your Business.* This was the original motto on U.S. money—it was changed to "In God We Trust" during the 1860s. Your business is your career-life. Pay attention to it, including the constant changes that need to be made in order to keep you and your business competitive.

Observations

- ❖ What is an intern/co-op? An engineer? No, you haven't had a chance to learn from your mistakes and be professional. A student? No, we expect you to be "almost" an engineer. Act like an engineer and contribute to your full potential at this point.
- ❖ Engineer—conceptual overview: An engineer applies the principles of science to solve practical problems and build products or processes. That's very different from a scientist who studies, predicts, and evaluates scientific events.
- ❖ Your ABET-accredited engineering curriculum provides you with a high-quality education that balances engineering theory with practical knowledge. Your education allows you to apply principles of science to build things and to solve engineering problems. You need a blend of both the knowledge of engineering fundamentals and the skill to apply those fundamentals. Strength in theory and strength in application: that's a balanced engineering education.

What Does Industry Expect of You?

- ❖ Be an interested, friendly individual who can contribute immediately to the company's progress.
- ❖ Top-level math skills.
- ❖ Be on time, properly dressed, with a positive attitude.
- ❖ Don't be afraid to get your hands dirty. Go without hesitation into the lab, machine shop, and so on.
- ❖ Engineers make things and fix things. As an intern/co-op or new graduate you can and should do this. Learn from your involvement and hands-on experience.
- ❖ Quoting Ben Kujawinski, production manager and former intern, "Remember, you only have one chance to make a good first impression."

What Can You Expect from Industry?

- ❖ Tips that will help you transition from the academic to business world.
- ❖ A revised work ethic—it's a different way of life.
- ❖ Constant new challenges at a fast pace.
- ❖ Cost consciousness with global competition.
- ❖ Quality consciousness—it will be instilled at almost a spiritual level.

Points to Remember as You Approach Your Professional Future

- ❖ You have a superior, ABET-accredited academic background.
- ❖ Your youth, enthusiasm, and academic knowledge are your biggest assets. Be generous and let others use them.
- ❖ The body of engineering knowledge turns over at least once per decade. Therefore, prepare today for the position you want in 10 years.

CHRISTINE B. WHITMAN

Ms. Whitman is the Chairman and CEO of Complemar Partners, Inc., a packaging and fulfillment company. She is also the Managing Partner in CSW Equity Partners, where she manages a portfolio of equity investments. From 1990 to 2000, she served as Chairman, President, and CEO of CVC, Inc., a worldwide supplier of thin film process equipment used in the manufacture of magnetic recording heads for disk drives, advanced semiconductor devices, and optical components. Ms. Whitman also serves as Chairman of the Board of Soleo Communications, Inc. and OneStream Networks.

Ms. Whitman has spent most of her career as an engineering executive. She began her career doing biochemistry research. She then joined a small company in the emerging electronics industry as a technical product manager. The company developed and manufactured thin film process equipment for the semiconductor and related industries. Ms. Whitman decided to put an investment group together and acquired the company, CVC, Inc., when it was about a $10 million/year operation. Ms. Whitman became President and CEO and built CVC to a much larger company that she headed for 10 years.

During the 1990s, new technology was fueling advances in the electronics industry. CVC focused on developing specialized thin film processes for next-generation devices. CVC employees worked on successfully winning one customer at a time to the point that they achieved leadership for supplying thin film deposition equipment used to manufacture the heads in disk drives. This involved process engineering, materials science, mechanical engineering, and control systems engineering. The market was moving so rapidly that they had to figure out how to develop large, complex machines, then manufacture, test, and ship them in less than four months, while producing them with a high level of reliability. In order to determine what products the company should develop, CVC partnered with their customers to identify competitive next-generation technologies. They assembled teams of outstanding mechanical and electrical engineers, physicists, and material scientists who loved building things and inventing innovative solutions. An especially important lesson they learned was to thoroughly understand and deliver exactly what each customer required. In business, every great idea needs a customer willing to apply the idea and pay for the product.

Now, here are some lessons that Ms. Whitman would like to share with you that helped her company achieve success.

Learning Along the Way

- Use experimental design to bring products to customers quickly.
- Consult with experts in your field to find solutions to your problems. These experts may be within your organization, your university, or a technical society. Join technical societies and find mentors. A mentor will be glad to tell you her story. This will accelerate your learning curve. Ms. Whitman learned about vacuum technology by contacting those who invented all the different components; they were very helpful.

- ❖ Understand the whole picture of your organization. Understand how projects will pay for themselves and achieve a return on investment (ROI). Understand accounting and work to stay within budget.
- ❖ Improve your communication skills, both written and oral. Most people dislike public speaking, but get past the fear and develop confidence and competence as a public speaker.
- ❖ Quality is critically important. Know about total quality management and ISO 9000. Customers expect products to run and operate smoothly. Whether it's an airplane, a car, or an appliance—we all expect products to work again and again. Customer satisfaction is, in large measure, determined by the quality of the products and services that are delivered.
- ❖ To be competitive today, focus on efficiency and fast time-to-market. New product development, project management, and marketing need to work together quickly and efficiently in order to bring novel concepts to the market before the competition. Planning, leadership, and teamwork are important at all levels.
- ❖ Know the ROI. Always know the value your project will deliver and make it happen. Understand, as much as possible, how your work fits into your organization's goals.

When You Get to Your Job

- ❖ Show up on time, dressed appropriately, use good manners, and be organized.
- ❖ Stay connected with your university for networking, including professors interested in your areas of technology.
- ❖ Join technical societies. Volunteer. Stay involved—hang out with others who share your technical interests. In order to stay on your technical career track, stay current. Know what's going on at the cutting edge of your technology and remain curious and passionate.
- ❖ If you're in research, development, or design, practice a disciplined approach—keep lab notebooks. File for patents. Build your intellectual property (IP) portfolio.
- ❖ Connect unique technology innovation to solve customer problems in novel, cheaper, faster, better ways. Think of your own buying habits.
- ❖ Deliver solutions flawlessly, on time, and better than anybody else.
- ❖ Start projects with a requirements document. This involves meeting with the customer (internal or external). Then deliver more than you have committed.
- ❖ Demonstrate feasibility. Convince your customer to test the product, or partner with the government to test it. Your technical concept usually won't work the first time. Therefore, try to get other people to pay for testing.
- ❖ Follow a sensible new development process. Specifically, validate your competitive advantage. Make sure that what you're building is better than what's out there already.

- ❖ Have an obsolescence plan—plan for the next generation of your product.
- ❖ Recognize what you don't know. And don't be afraid to work with someone who's smarter than you.

What to Expect

- ❖ There will be more work than you can possibly do.
- ❖ There is never enough money in the budget.
- ❖ No one will tell you exactly what to do.
- ❖ Customers don't always know exactly what they want.
- ❖ Priorities change constantly.
- ❖ There will be a crisis regularly. Sales will sell something not yet developed and you'll be assigned to deliver it immediately.
- ❖ Find a way to increase revenue and decrease cost. You will be appreciated and rewarded.

Advice

- ❖ Develop a life plan. Define your success.
  - What are the most important things in your life?
  - What are your skills, interests, and values?
  - What activities do you enjoy most?
  - What is your ideal work environment?
  - What can you do that needs changing in the world?
  - How do you want to be remembered?
  - What do you feel compelled to do?
- ❖ Translate your dream into an action plan.
  - Putting your plan in writing provides a roadmap to achieve your dreams.
  - There's plenty of energy when your work is enjoyable.
  - If you can't get through the day, you're in the wrong job and need to do something about it.
  - Manage your fears with your action plan.
- ❖ Things work out, but plan on regular doses of setbacks in the way of embarrassments and failures. Setbacks keep you humble and help you get better.
- ❖ Don't blame others—your parents, lousy boss, partner, poor teacher, or coworker. Blaming wastes valuable time, energy, and attitude. Stay focused and get the project done.
- ❖ Always listen to your customer.
- ❖ Get and give help along the way.
  - Join volunteer organizations. Contribute for the greater good. Meet fascinating people and learn what different organizations do best. Benchmark others. Learn examples of best practices in all walks of life. Read success stories. Ask people about their successes.
- ❖ Anticipate problems—more things go wrong than you can imagine. So, under-commit and over-deliver.

Two Actual Engineering Co-Op Stories—a Winner and a Loser

1. *Great Co-Op Story (Winner).* A co-op student had a project goal of figuring out how to create a unique thin film process. While experimenting, he successfully created a unique solution. Many brilliant engineers had worked on the problem; however, no one had thought of this idea. The device was patented and became a very successful product.
2. *Poor Co-Op Story (Loser).* A co-op student was working with a customer who had flown in from Europe to test and accept a multi-million dollar machine. The co-op student left the customer and an important test to go see a movie with a friend. As a result, he missed the opportunity to travel to Europe to assist with the start-up of the equipment and, ultimately, a great job offer.

Summary Tips

- ❖ Get passionate about something and become an expert. Be the best. People will need your talent.
- ❖ Begin your action plan to be financially secure now. Save one year's salary and keep it in the bank so that you can be independent and take more risks—you can quit your job if you don't like it.
- ❖ Study best practices, read regularly, and learn from those who have succeeded.
- ❖ Stop doing things that aren't helping you succeed. Change your habits accordingly.
- ❖ Appreciate your most important assets—your family and friends. They'll be the ones who will be there for you and you for them.
- ❖ Follow your principles—your actions define your integrity. What you do is who you are.

LAWRENCE L. PECKHAM

Mr. Peckham is the founder and retired chair of LPA/Xelus Software, Inc. He holds a BS degree in Industrial Engineering. After several years of experience at Xerox, Mr. Peckham founded LPA Software while obtaining his MBA. LPA/Xelus Software, now an affiliate of Illinois Tool Works, Inc., is now the world leader in inventory planning software for high-tech, field service and maintenance, repair, and overhaul (MRO) markets. Mr. Peckham has received numerous honors and distinguished engineering awards, including Business Person of the Year by *Business Strategies* Magazine.

Mr. Peckham began his career at Xerox as a distribution engineer in a department that designed and installed warehouses throughout the world. To increase his efficiency, he got a Fortran programming book, taught himself to program, and started writing computer programs. One result was that calculations his boss expected that would take two weeks to complete, Mr. Peckham produced in seven seconds through the computer.

Within two years, Mr. Peckham's computer analysis concluded that the job that he and the department were doing shouldn't be done—they should not be building warehouses any more. So, he took a bold step by convincing his boss and upper management that they should not be in the warehouse-building business. The result: the decision was made to dissolve the department. The only one left was Mr. Peckham who was given the request to keep programming and coming up with new revenue-enhancing ideas for Xerox. Mr. Peckham learned some important lessons from this that he'd like to share.

- ❖ You want to be on the revenue side of business, since the ideal cost center is zero.
- ❖ When you're an engineer, you are in the business of cutting costs. This may sound scary, but since the ideal cost is zero, you should be trying to eliminate your job. Come up with a new idea to do that along with an approach that enhances revenue and you'll be rewarded, not fired.
- ❖ Find out what you're good at and what you really like to do, then join a company where your talents are at the core—the center of their business.

Mr. Peckham decided to earn his MBA. Meanwhile, Xerox said, "You're such a good programmer, why don't you continue working for us in your apartment while going to school and we'll pay you." He worked, they paid, and it turned out that he was making more money programming Friday through Sunday than he had been as a full-time employee at Xerox. So he went into business for himself, founding LPA Software Inc.

For the first five years LPA Software, Inc. had one employee—himself. Eventually, business success required him to hire another employee. By 1985, LPA Software had 12 employees with sales of $900,000 per year. Then they developed a written plan to grow 25% annually for 12 years. The plan turned out to be accurate—they met their objectives within 5% each year, and by 1997, sales grew to $24 million with 25 employees. In 1999, Mr. Peckham sold his interest in LPA at a $78 million valuation, retiring at age 52 to pursue other interests.

At this point in his story, Mr. Peckham appropriately invites you to play "Who wants to be a millionaire?" by answering the following questions.

Who Wants to Be a Millionaire?

1. A salesperson is:

a. Not me—I am an engineer.
b. One of life's worst nightmares.
c. Something I would like to be.
d. A necessary evil.

Mr. Peckham invites a class of engineering students to vote, by raising their hands, as to which of the above choices they personally believe to be true. Think about it. What's your answer? In an average group of engineers, "c) something I would like to be" gets the fewest votes. But Mr. Peckham believes that "c" is the best answer for successful engineers, whether or not they are in an official sales job. His reasoning is that engineers, like everyone who is successful, need to sell themselves. You have to sell yourself to get a job. You have to sell your ideas to get your project funded. Indeed, the highest-paid individual contributors in a corporation are generally the salespeople. They can make up to $400,000 per year. A technically skilled person, especially an engineer who can sell and wants to sell a technical product, is extremely valuable and has great potential for success. But whether or not you are officially in a sales function, stay close to customers and recognize that listening to customers and selling yourself and your ideas will always be keys to your career future.

2. Regarding communication in a business group, it is best to:

a. Share your ideas openly and often.
b. Be quiet until cornered for your input.
c. Comment, but only when you have all the facts.
d. Listen, until you have experience equal with others in the group.

What do you think? Mr. Peckham feels that "a) share your ideas openly and often" is the best strategy. You have a lot to offer. Don't doubt yourself, especially if you have an idea that you believe is worth sharing with the group. Engineers, especially new engineers, could have a tendency to comment only when they have all the facts while listening until they have experience equal with others in the group. Regarding item "c," consider this—knowledge is power, but if you wait until all the facts are in, you may miss the opportunity to make a valuable contribution or it may never even happen. This is also true of choice "d"—it is very important to listen, but you may never have experience equal to all the others in the group. Emphatically, listening is very important; if you're not willing to learn and change, you're not listening. Listen well and speak intelligently when you have a good idea worth offering.

3. If, on your job, you don't know what to do, you should:

a. Not do anything—you might do something wrong.
b. Wait for clearer guidance.
c. Ask for help and demand action.
d. Do anything. Doing anything is better than doing nothing.

Mr. Peckham believes that "c) ask for help and demand action" (without being too rude) is the best answer on this one. Choices "a" and "b" would have you passively waiting for help that might never come. As a co-op/intern or as a new engineer, you certainly don't want to come to the end of your assignment or appraisal period with nothing accomplished since "no one told me what to do." Choice "d" would have you spinning your wheels and producing nothing, perhaps even getting in the way and being counterproductive.

4. Compensation:

a. New hires are never paid what they are worth.
b. Changing jobs frequently leads to the highest long-run compensation.
c. Compensation may not be fair in the short run, but it is always fair in the long run.
d. It is important to earn more than your associates.

Again, Mr. Peckham believes that "c" is the best answer. He especially believes that compensation is always fair in the long run. Either you will make it fair or the company will. In the meantime, don't worry about what the person next to you is making. Focus on your own outstanding performance and outstanding pay will follow.

5. **The value of an engineering co-op/internship job is:**
   a. The money.
   b. The opportunity to learn new skills and experience business in action.
   c. The opportunity to show the world what you can do and to build your reputation.
   d. The opportunity to test what you've learned in class.

   All of these values are important. As a student, you have an immediate need for money to live and pay for your education. Your compensation will help you take care of that. More important will be "b" and "d": your opportunity to apply what you've learned in class, to learn new skills, and to experience business in action. This combination will result in item "c"—the opportunity to show the world what you can do and to build your reputation. Specifically, during your co-op/internship assignment, you should have the goal of producing value for your employer so that you can put quantifiable evidence of that on your resume. Your valuable contribution as a co-op or student intern is likely to result in a job offer when you graduate.

KITTY PILARZ

Ms. Pilarz is Director of Worldwide Product Safety for Mattel, Inc. She holds a BS degree in Mechanical Engineering and an MBA degree. She is Chair of the Mattel/Fisher-Price Safety Committee and is a leader on numerous committees for the American Society for Testing and Materials (ASTM), the American National Standards Institute (ANSI), and Underwriters Laboratory (UL). Ms. Pilarz is also a board member and past president of the International Consumer Product Health and Safety Organization (ICPHSO).

Ms. Pilarz is having a successful career that she has found very gratifying. She was not sure it was going to turn out this way. She really enjoyed being an engineering student, but suspected the fun might be over when she entered industry. Her career turned out much better than she anticipated.

Therefore, as a professional engineer and engineering executive, Ms. Pilarz is motivated to share her thoughts with you.

How to Make Engineering the Profession You Want It to Be

- Be passionate about what you do. Choose an industry and an employer that you believe in. Be proud of the product that you engineer. If you accept a job just for the money or convenience, it may be very difficult to reach the level of passion and enthusiasm that will make your job a joy. That would especially be the case if your company is making a product that opposes your values. There are many opportunities, employers, products, and functions out there. Choose something that can ignite real passion in you each day.

- Always do your homework. As worldwide director of product safety for Mattel/Fisher-Price toys, Ms. Pilarz is constantly conducting research on the latest information on child safety related to her company's products. She finds it challenging to be at the cutting edge of research as to what the curiosity and imagination of a child will do with a particular toy. Will a child use two toy buses as roller skates? Will a little girl be so fascinated with her dollhouse that she wants to put her head through the door? These are engineering safety issues for child customers, who are too young to be thinking about safety. That could be scary, but it's also fascinating, and Ms. Pilarz loves the opportunity to ensure that Mattel/Fisher-Price is making safe toys that children will enjoy.
- Get all the training you can. That can certainly be technical training. It can be communication and management training. In Ms. Pilarz's case, it also means media training—how to interact with aggressive news media people during a potentially negative situation. Media training was difficult, but Ms. Pilarz is glad she did it. Training keeps you sharp, allows you to grow professionally, and opens new areas of opportunity.
- Be international. In our global economy, there is a career advantage to having international experience. That can mean international assignments or travel opportunities. If possible, develop international language skills. Having global perspective and experience makes you a more eligible candidate for promotion.
- Volunteer. Successful people do volunteer work. Volunteering your time and talent within your company or community pays off. You'll probably be volunteering for something worthwhile that you're good at and enjoy. It allows you to have the satisfaction of making a valuable contribution, networking with others, and enhancing your professional visibility.
- Be a problem solver. Don't just say we have a problem—propose a solution! Many people can spot a problem. Some say, "we ought to work on this," and too few propose a solution to the problem. Be the first one to put a suggested solution in writing. Problem solvers get promoted.
- Challenge yourself as much as possible—for career growth and to make the job interesting. For example, Ms. Pilarz had to set policy and design standards for when products would require children to wear helmets, knee pads, or other safety protection. This meant doing research on how fast a child could go on roller skates, bikes, and so on. The research involved unique and carefully orchestrated races of several hundred children on toy vehicles. It was a technical and organizational challenge, but it was also fun and produced valuable results.
- Choose an industry that you believe in and a function in which you can excel. Find your passion. This will allow you to challenge yourself and keep growing while having fun with your work. You can do these things which will contribute to a fulfilled and meaningful life.

SHAWN THOMPSON

Mr. Thompson is Director of Engineering for Sigma International, a manufacturer of state-of-the-art medical devices. He holds a Masters degree in Mechanical Engineering. Mr. Thompson has broad experience in development, design, and engineering management in several engineering companies. He began as a co-op student and started his career in the automotive industry. Subsequent experience included technical and management assignments with Leica Inc. (diagnostic instruments and optical microscope technology), then Strippitt, Inc. (electronics and controls for large machine tools).

Mr. Thompson draws from his extensive knowledge of engineering and business management as he shares his perspective with you.

How to Make the Transition from Academia to Industry

- First, recognize the difference when you move from campus to industry:

As an Engineering Student	As an Engineer in Industry
You are a paying customer.	You are a paid provider.
Focus is on your individual performance.	Focus is on company performance.
Focus is on testing and grades.	Focus is on production.
Emphasis is on theoretical knowledge.	Emphasis is on practical application.
Your environment is local.	Your environment is global.
Your communication is personal.	Your communication is interpersonal.

- What is engineering? A lot of things. It's thinking. It's solving problems. It's analyzing and designing. It's tinkering, building, and inventing. It's communicating and selling your ideas. It's also planning, working with customers, and participating in meetings. Depending on your function, it can require any or all of these skills. The core of engineering is technical, and the pace of technical change grows faster each year. You need to keep up with your area of technology to be competitive. But there is another big dimension to engineering, and that dimension is business.

The Business of Engineering

- Why should I hire and pay an engineer? That is, why should I hire you? Answer: to add value and make money. And that leads to the next larger question.
- Why is any company in business? To make money by providing goods and services. Hopefully, you and your products are making the world a better place, but if your company is not making money, it is no longer in business.

- Three things are needed to produce goods in the manufacturing or service sector.
  - Manpower—people to do the work
  - Materials—physical things to perform work upon
  - Machines—tools and equipment
- Money made from goods: profit equals selling price minus cost of goods.
  - Selling price is determined by the market—you don't get to pick it. Selling price is determined by your customer or competition; it is what your customer is willing to pay.
  - The cost of goods is the only controllable value or variable.
  - In industry, profit is the driving force, but cost is the controlling force.
- Visible cost of creating goods includes research and development, design, manufacturing, and marketing your product. It includes paying your salary. Visible costs are those you can foresee and budget.
- Hidden costs of creating goods—include non-optimum design, design risks, unexpected overhead costs, manufacturing variance, low-volume production, regulatory costs, and legal costs. Non-optimum design includes prototype development and the fact that 80% of a product's cost is within the first 20% of its life cycle. Hidden costs are those that, by definition, are difficult or impossible to anticipate. Nonetheless, hidden costs are real and need to be minimized, and if possible, turned into visible costs that can be better managed and reduced.

You—The People Side of Engineering

Engineering is a business investment that creates value. Now that we know about profit, costs, and where costs come from . . . where can you as an engineer create value? You are a business investment. Your engineering success will be defined by your ROI factor. Therefore, you want to:

- Maximize the return on investment (ROI) you are providing to your employer.
  - Be aware of the business of engineering. It may sound callous, but you're in it for the money. Your employer feels the same way about having you on their payroll.
  - Be aware of the fact that you are in a cost-driven industry. Do it right. Minimize cost, but not to the point of unacceptable quality.
  - Value engineer everything you do. Value engineering evaluates each step in the process of creating a quality product at minimum cost. The importance of value engineering is considerable and is covered separately in Chapter 13.
  - Sell your value that you bring to the company. This requires communication with your management and customers. In hard times, this will determine who your company can afford to pay and keep. At all times, it will determine who has the most potential for assignment to an important project and for promotion. It's all

about how your company perceives its ROI by employing you and in what capacity.

- Be confident of your ability and potential. You will greatly increase your probability of engineering career success if you have a plan, remain competent, and sell your value to your customer and employer.

PANEL DISCUSSION

The following are questions posed by students and our executives' responses.

1. **What is it about a resume that will make you interested?**

   A: Put yourself in the position of the person reading the resume. Give them the best, focused information relevant to that employer's need. Be honest, to the point, and neat. Spelling and grammar are important.

   A: Have your objective match the job requirement. Summarize experience and accomplishments in your job history and other interests. There is an advantage to being well rounded in life beyond your expertise.

   A: I go through several elimination steps: first I look for the engineering major that I seek. Then I look for a cover letter that is succinct, well written, and with no grammar or spelling errors. Then I grade each resume based on what I'm looking for. The best resumes are well formatted, concise, and results oriented, with no spelling errors.

2. **What are your selection thoughts about who you hire? How can I make the best impression?**

   A: I want to hire the best. Skills, personality, and work ethic all combine to make a good impression. Also, you need to want the job. Some people coming out of school don't seem to want the job. They lack apparent motivation. Some people seem to expect to be hired and don't anticipate that they're going to have to work very hard. Show me that you're competent, motivated, and really want to work for me—that goes a long way.

   A: Besides how to get hired, you also want to decide whether you want to work for that company. Does that company fit with your life goals? Look into a company's guiding principles or code of ethics. Do they have good values that correspond with yours? You have a tremendous amount of talent that you're bringing to the transaction. You want to select the company that gives you the most opportunity. It goes both ways.

3. **What characteristics do you look for in a good employee?**

   A: I have a job opening. It makes sense to pick the person who wants the job and is best able to do it. Make sure there is mutual understanding about what needs to be done.

   A: Initiative. After being given a job by their boss, it's great when a new employee can take the initiative and soon see what needs to be done and then take action. Ideally, the employee will write up what needs to be done, clear it with the manager, and go do it. This quality is especially valuable in an organization where everyone is so busy that they don't have time to define exactly what's needed of a new employee.

A: I like to see engineers willing to sit down with customers to create opportunities and work through problems. That directly enhances revenue and cuts cost rather than cycling information from one department to another. Get information and approval as needed and make the deal or solve the problem.

A: In contrast, employees who are grumpy and unhappy are likely to fail. Why? Because they probably don't love the work and are probably in the wrong job, but may not want to admit it. We may let such people go if there's nothing else in the organization for them. In that case, we're doing them a favor because they've been in the wrong company and in the wrong job. They would be better off somewhere else, and they need to figure out where that is.

4. During your first week on the job, how do you make a good impression?

A: During the first week, meet with your boss and get a handle on what you're supposed to do. Write it down, including how you're going to accomplish that.

A: I agree. Make sure you meet with your boss and get understanding and acceptance in your action plan for what you need to do. It's key that you know what needs to be done and your boss agrees with you. Then, be sure you understand the resources (people) you need to work with to get the job done and then meet them. There are always internal customers or suppliers who can help you and work with you. Know your information technology (IT) person.

A: Show up on time, be friendly, work hard, and stay late.

5. Tell us about a difficult challenge on your first job.

A: A marketing guy put me into a meeting with an irate customer. That customer was angry. I winged it: I listened to his complaints and wrote them down. Then I apologized and told him I'd try to help, which I did. I took some positive action, cooled the customer down, and meanwhile learned a few things about solving problems.

A: I was concerned when I agreed to work in a marketing organization with a purely technical background. It turned out well though. I learned that if you don't have all the skills and experience, there are others around you who can help. And they'll be willing to help. With the right attitude, put your mind to it and you'll do a fabulous job.

A: I was scared about what I saw as unrealistic expectations of me. My advice to you is to persevere. If you really want to do well and you're committed, put in a serious effort; there is a high probability you'll be successful. On the other hand, if you're lazy and sit back, you won't make it.

6. **How much do you value versatility in an employee?**

   A: Design, writing proposals, testing, and production all require versatility, especially if your job crosses more than one of these functions. Versatile generalists are needed. A company also may need some very specialized people where versatility is not that important. Coming out of school though, especially at the BS level, you don't want to become a narrow specialist anytime soon.

   A: Sometimes a job requires more versatility than indicated in the job description. You may be dealing with a lot of different people, departments, technologies, and customers. So, versatility is important from a technical, communication, and business standpoint. If you're successful, you will be asked to move to different functions and perhaps different locations.

7. **How beneficial has an MBA been in your career?**

   A: The one thing lacking in my engineering education was finance. My MBA has allowed me to make better financial business decisions.

   A: It was tremendously valuable. It was especially important to gain some experience rather than going into an MBA program right after my engineering degree. With experience, you can grasp the reality of the MBA course information.

8. **How soon after getting your engineering degree(s) should you consider enrolling in an MBA program?**

   A: The longer you wait, the harder it is. I started my MBA part-time after starting work and the company paid for it. The idea of waiting 10 years, then deciding to get an MBA creates problems—especially if you're then married, and have children and business/travel obligations. It's hard to do it all.

   A: I got my BS degree, then a year later went back at night. Getting an advanced degree at night is a real personal commitment. Now, with four kids, I wouldn't be able to do it. I must add that I struggle with the idea of hiring a person with an MBA right after obtaining their BS in engineering without experience. I'm not sure whether they really want to be an engineer.

   A: It's best to get your MS or MBA within a few years out. It all depends on your goals and situation, such as, whether or not you have a family.

9. **If you just have your BS in engineering, are you going to be able to make it into management?**

   A: It all depends on what you're doing and what your goals are. I prefer an MS in engineering. I can teach an engineer management and finance, but not necessarily the newest technical subject matter at the graduate level.

A: A misconception by some is that if you get a BS in engineering, then immediately an MBA, you might be able to move right into management. The fact is, no one is going to give you an engineering management job without functional engineering experience.

10. How important is it to get a Professional Engineer (PE) license?

A: If you're a civil engineer, it's an absolute necessity. I don't think it's as generally important in other disciplines.

A: In certain functional areas, you have to have the credentials to defend your professional integrity if challenged. By that, I mean I represent the company in litigation. One of the things the plaintiff's attorney will do is to attempt to make it appear that I don't have the technical chops for the job. If you are in that situation, you'll be much better off to be able to prove that you have all the appropriate certifications and licenses.

A: If you are working on a government or public works project, you are likely to have to sign drawings and have them stamped with your PE license.

11. What's your suggestion if I come up with a really good idea? What do I do?

A: It's best to have an idea you can make money with, or it won't go any further. Some people have had great ideas, but no one could figure out how they could make money with those ideas.

A: Be specific with your idea. What is it? Write it down and show how it can work and be profitable. If you come up with the idea first, you should patent it. If you want to see your idea quickly on store shelves, tell somebody else about it—someone who has a factory and is able to produce it. Suddenly, your idea is theirs and it's their product.

A: Whether it's a revenue enhancer or cost-cutting idea, it's really important to put it in writing and persuade your management that it's an idea worth pursuing. Patent it.

12. What are your thoughts about life balance? How do you balance your career with family and other things in life?

A: I suggest you write down a life plan, including your objectives. What's important to you? The plan can evolve—the more experience you have, the more aware you are of what's important, and the more input goes into the plan. You can balance the benefits versus the costs of achieving that plan. Only you can figure that out. But if you don't have a plan, you may drift around and not come up with the best results.

A: It's a challenge to devote a lot of time to your family and also be highly successful in business. It's about your values and choices. You may not want to go to an extreme in either direction. It's certainly possible to make a sensible balance that includes a good career, a good family life, and other interests. That does require organization and self-discipline.

Chapter Summary: Takeoff Tips

You have just read a valuable range of experience and advice from five top-level engineering executives. They have presented their stories with the goal of giving you helpful information for your career success. Here is a summary of their wisdom for your reference and action.

- ❖ Select an employer that matches your values and goals.
- ❖ Choose work that can ignite a real passion in you each day.
- ❖ If a company is not making money, it is no longer in business.
- ❖ You will be hired to add value and make money for your employer.
- ❖ Want the job.
- ❖ You will always be in sales. Sell the value that you bring to your employer.
- ❖ Skills, personality, and work ethic all combine to make a good first impression.
- ❖ Meet with your boss and agree on your written project action plan (Chapter 10).
- ❖ Cost, quality, and schedule are key requirements for your project (Chapter 12).
- ❖ Practice good soft skills of communication, punctuality, networking, and so on.
- ❖ Be versatile and able to work with all functions, including non-engineering.
- ❖ Understand the big picture of your organization, including return on investment.
- ❖ Maximize the return on investment you're providing to your employer.
- ❖ Be customer focused; your customer is key. Understand the customer's needs.
- ❖ Give help and get help.
- ❖ Value engineer everything you do (Chapter 13).
- ❖ Challenge yourself. Have initiative and take action (Chapters 20 and 21).
- ❖ Be a problem solver—don't just define a problem, propose a solution.
- ❖ Be international (Chapter 19).
- ❖ Technology changes fast. Keep up (Chapters 11 and 18).
- ❖ Develop goals and write a life plan. Balance your professional life and personal life (Chapters 3 and 20).

Exercises

1. You have read success wisdom from three CEOs and two engineering directors. What did you learn from:
   a. Mr. Taylor?
   b. Ms. Whitman?
   c. Mr. Peckham?
   d. Ms. Pilarz?
   e. Mr. Thompson?

2. Who especially influenced you to act on his or her advice? What was the advice and how will you act on it?
3. What specific things did you learn from the panel discussion?

Looking Ahead

Throughout the rest of Parts 2 and 3 of *Ready for Takeoff!* you will read specific, functional wisdom from experts about how you can apply the executives' advice from this chapter on the job. You will obtain knowledge throughout the remainder of this book that parallels and reinforces what these engineering executives have described. In subsequent chapters you will learn more detailed, application-oriented information about what it takes to be a total engineering professional. Each chapter is important in preparing you to be a "most qualified" employment candidate and new-hire.

Now, let's proceed, in the next chapter, to project management. What is it, and how do we manage a project?

CHAPTER 12

Project Management

You, as an engineering student, are already a project manager. Who me, you say? Yes, you are managing the project of getting a BS, MS, or PhD degree in engineering within X number of years at Y cost with a targeted grade average and, desirably, some technical co-op/internship or research experience before you graduate. Moreover, you are managing the project that defines this book: the launching of your career that optimizes your skills, interests, values, and employment market need. Also, as a project management objective, let's say you'll decide to nail down the right job for you by commencement. Project management is all about defining such goals and making them happen. Learning and transforming project management skills into habits will be a tremendously positive step toward your professional and personal success.

Now, presume you have the job. Picture yourself on your first co-op/internship assignment or just starting out as a new graduate. You meet with your supervisor, get oriented, and are given an assignment. You guessed it: your assignment is a project. That means you're responsible for managing yourself, and perhaps others, to use your technical and personal skills to get results. That's why you were hired, remember? You were hired to be an asset and create value; this is your opportunity to make a great first impression and accomplish a job as specified, within time and budget limits.

Project management is of sufficient importance that joining in this chapter are two outstanding project managers, Bill Jehle and Blair Webster, who will share their expertise for your advantage.

Bill Jehle received his BS degree in Civil Engineering from the University of Notre Dame. He spent much of his career as a project manager of teams building multi-million dollar cryogenic air separation plants. Blair Webster received his BS degree in Electrical Engineering from Northwestern University. His experience is mainly in software project management.

DEFINITION OF PROJECT MANAGEMENT

Bill Jehle, civil engineer and veteran project manager, defines a project as follows:

- ❖ A project is a planned undertaking with a well-defined beginning and end.
- ❖ A project has a schedule, budget, and built-in measures and controls to ensure a successful outcome.

Then, Mr. Jehle explains that the three most important things about a project are definition, definition, and definition! If you can't define what you're going to do, how do you know where you're going or how you're doing? A project can be anything, but you need to define its boundaries, know the beginning and end, and know where you are at any point in time. You need to know what's in and what's out. You need to plan, organize, lead, and control. You can do a great job on the technology, but if there's a cost or schedule overrun, you're in hot water. What are the consequences of delay? Are there penalty clauses in the contract? Technical features, cost, and schedule need to be balanced on an engineering project to ensure that the customer has what is needed, when it is needed, and at the level of quality agreed upon.

WHO ARE THE KEY PEOPLE IN AN ENGINEERING PROJECT?

Project Sponsor

The person or group who champions the project and provides the money for its success is the project sponsor. The sponsor is likely to be an important person within or outside your organization. Sponsor support is vital to the project's success. Therefore, the sponsor needs to be informed of the project's definition, progress, and any problems.

Project Stakeholders

The people who use the results of your project are the project stakeholders. These are your customers who need to be satisfied. You need someone within your organization and the customer's organization who can effectively communicate on technical and business matters as the project progresses.

Project Manager

The project manager is the one who completes the task of setting objectives, planning, organizing, leading, and adapting to change. The project manager is ever conscious of the necessities of delivering agreed-upon technical results at cost and on schedule. If the customer, part-way through the project, desires a technical enhancement that will increase cost and delay schedule, that change needs to be agreed upon by all key people, and that the project definition must be adjusted accordingly in writing. As Mr. Jehle explains, the project manager should always display a confident and positive posture to people associated with the project. Give away any praise or recognition to the team and take personal responsibility for any problems or shortcomings. As project manager, you will be amazed at how much credit you will personally receive for trying to give all the credit to the team.

Project Team

Most complex projects will require the skills of a diverse project team to accomplish the defined project's technical objectives. Members of the project team generally do not permanently report to the project manager. Team members may come from functional departments relevant to the project, including sales, estimating, purchasing, manufacturing, and relevant engineering design functions such as process, mechanical, electrical, or civil. The project manager, as team leader, must serve the roles of coach, mentor, motivator, and cheerleader for the team with constant focus on the final-stated project objectives. The team leader must recognize the strengths of each team member, meanwhile recognizing any weaknesses that need bolstering. Effective project team leaders do this to maximize individual talent while compensating as necessary to fully achieve the final objectives. Mr. Jehle's final team leadership advice: when things get difficult, the project manager should shield his team members. As a project manager, you might need those team members for your next project.

HOW DO YOU START A PROJECT?

- ❖ *Summary Statement of Problem or Opportunity.* Write this on a single sheet as an executive summary to your sponsor, explaining why the sponsor should invest attention and financial resources to support your proposed project. Definition!
- ❖ *Detailed Action Plan.* Be specific. Include everything you plan to deliver, at what cost and schedule. Specify what's in and what's out. Definition!
- ❖ *People and Equipment Needed.* Be sure your sponsor and stakeholders know what's required to achieve your goal. Definition!
- ❖ *Signed Agreement.* As project manager, be sure your sponsor, stakeholders, and project team agree on the broad details as well as specific details of the project. That's essential.

PROJECT MANAGEMENT TIPS FOR YOUNG ENGINEERS

Blair Webster, electrical engineer and another outstanding project manager, offers these useful tips.

- ❖ You're a resource. You will be put on project teams and will be accountable for your tasks. You are given deadlines and others may be depending on your work.
- ❖ Know that being a project team member is beneficial; projects get done faster, more efficiently, and more reliably. That means the right projects get worked on and there's more information available about project status.

- There's always a schedule. Someone is always responsible for the schedule, and that person is the project manager, whatever their job title really is.
- Get it in writing. Get a signed/approved scope of work before you invest any time in design.
- The first one to write something down wins. If you're in a meeting and are the only one writing details of the discussion, you can expect to get 80% of what you express in the meeting minutes you present to the team. Others will do minor editing, but most of what you write, presuming you are accurate and responsible, will stay intact.
- Projects have a triangle of technical features, schedule, and cost. A change in one means a change in the others. You can't change one side of a triangle without affecting the others. If you're building a house, then want an extra bedroom, it will affect the cost and schedule.
- Prepare for meetings; your engineering skills will be judged on your communication skills. This may sound unfair, but other people on your team aren't watching you as you perform your calculations and individual work. What they see in the meeting is whether you're organized, well prepared, and well spoken. Take 15 minutes before a meeting, get your notes together and figure out what you're going to say. Present well and convincingly on a technical subject and you will be judged to be a good engineer.
- "Almost complete" is a schedule killer. Don't do it, and don't let others do it to you. There is a tendency by some people to casually and grossly underestimate the time remaining on their part of a project. If it's almost done, be sure it's ready for testing.
- Know who your customer is. Who will be accepting or testing the results of your project? What are their criteria for your success? If you're working on software and your customer just cares what it looks like and that it is easy to use, you're wasting your time making it faster. Talk to your customer and ensure that your work meets their specifications—that is, the intended use and need.
- Change happens; manage it effectively. Change in input means change in output. Raise your hand with delays as soon as they're discovered. You may get a task in writing, then the marketing person says the customer wants a new feature; engineering says ok, but doesn't tell marketing that it will take another month. Bad news—in a situation like this, don't let engineering be you. The right approach is to say, "We can add that, but it will take another month. Is that ok?" If so, it needs to be put in writing and communicated to the project team.
- Don't underestimate non-technical aspects of a project. It is essential for a successful project team to trust each other and communicate openly.
- You will be involved in teams that use project management techniques. The team's success will be your success. The more you know about project management, the more you can contribute to your team's success.

FINAL WORDS

This information about project management has been provided by seasoned project management experts, Bill Jehle and Blair Webster. It is a valuable summary and will provide you with what you need to know as you step into engineering project employment with industry or government. Much has been written about project management, but Mr. Jehle's and Mr. Webster's experience and advice put it into a nutshell for you to use. For more details, a well-recommended book on project management is *Making Things Happen—Mastering Project Management,* written by Scott Berkun, a Microsoft project manager. You can also become certified as a project management professional (PMP). Obtain more information about becoming a PMP by visiting the project management professional Web site at www.pmpcertification.net. Meanwhile, this chapter with Mr. Jehle's and Mr. Webster's advice will get you going with project management know-how as you enter employment as a project manager or project team member.

Chapter Summary: Takeoff Tips

Project management definition:

- A planned undertaking with a well-defined beginning and end.
- It has a schedule, a budget, and controls to ensure a successful outcome.

The key people in an engineering project:

- Project sponsor champions the project and provides the money.
- Project stakeholders use the results of your project.
- Project manager plans, organizes, and leads the project team.

Requirements for starting a project:

- Summary statement of problem or opportunity
- Detailed action plan, including cost and schedule requirements
- People and equipment needed
- Signed agreement to ensure sponsor stakeholders and project team agree

Project management tips:

- You will be put on project teams and will be accountable for your tasks.
- There's always a schedule.
- Get a signed/approved scope of work in writing before you design.
- The first one to write something down wins.
- Projects have a triangle of technical features, schedule, and cost.
- Your engineering skills will be judged by your communication skills.
- "Almost complete" is a schedule killer.
- Change happens; manage it effectively.
- Don't underestimate non-technical aspects of a project.
- The team's success will be your success.

Exercises

1. Why are you, an engineering student, already a project manager?
2. What, per Bill Jehle, are the three most important things about a project?
3. Name and describe the key people in an engineering project.
4. How do you start a project?
5. Name five of Blair Webster's project management tips that you'll use.

Looking Ahead

In Chapter 13 you will learn about value engineering. As you undertake a project, recall the advice of Engineering Director Mr. Shawn Thompson: "maximize the return on investment (ROI) you are providing to your employer. Your engineering success will be defined by your ROI factor." Therefore, "value engineer everything you do. Value engineering evaluates each step in the process of creating a quality product at minimum cost."

CHAPTER 13

Value Engineering

RICK LICURSI

Value Engineering (VE) is essential engineering knowledge. It is a privilege to have one of my former students, Rick Licursi, submit this chapter on the important subject of VE. I believe you will find this chapter especially valuable, since Rick not only explains the principles of Value Engineering and Value Analysis (VA), but also how he applied these principles on his undergraduate internship assignment with great results.

Rick is the Business Development Manager for the Computer Aided Engineering Center of Excellence at Siemens PLM Software. At Siemens, he works with high-technology companies of all sizes who wish to implement the latest in Computer Engineering Software. Rick holds a BS degree in Mechanical and Aerospace Engineering and an Executive MBA, both degrees from the University at Buffalo.

THE IMPORTANCE OF VALUE ENGINEERING AND VALUE ANALYSIS

Every engineer needs to know about and how to use VE and VA. They are extremely valuable techniques to ensure that, on the job, your engineering work includes steps to optimize intended product function at the lowest cost.

In this chapter I will first explain VE and VA. Then I will go into detail about how I applied these concepts on my engineering internship assignment, along with the specific results achieved.

A wise, old engineer told me to always remember these things: 1) keep things simple and 2) use your common sense. These are words that stuck with me throughout my college and engineering career. They are words that every engineer should remember when faced with the need to find a solution to any type of problem. I have applied many project implementation and problem solving methodologies over the years; but when used together, the two approaches that have never failed me are VE and VA. Not only do they conform to the simplicity and common sense criteria noted above, but they can be applied to any industrial sector, and are widely used in high-tech, government, construction, and service industries. Employing these methodologies is seen by many Fortune 500 companies as successful long-term business strategy.

WHAT IS VALUE ENGINEERING?

VE is a system that evaluates each step in design, materials selection, fabrication processes, and operations management so as to manufacture a product that conforms to its intended functions and has the lowest possible cost. The VE methodology was developed in the 1940s by an engineer from General Electric named Lawrence Miles. For more than 50 years, Lawrence Miles authored numerous papers that illustrated his passion for the promotion of VE. Why did he believe in it so strongly? For the simple reason that it is very important and different from traditional cost reduction techniques.

Traditional Cost Reduction Techniques	Value Engineering/Value Analysis
Limit focus on current cost	Detailed cost analysis
Part or feature oriented	Function oriented
Attack only the obvious	Attacks the ambiguous
Subjective in nature	Analytical in nature
Can generate excessive data	Provides detailed information
Can cheapen the part	Maintains function

CORE PRINCIPLE: ESTABLISH SUCCESS STEPS

One of the core principles of VE is establishing specific steps in the engineering, manufacturing, or distribution phases of product creation. Once this is done, a monetary value is established for the value of each phase of the product, which in turn reflects the functions and prestige value of the product. Ultimately, it is these things that make the ownership of a product desirable. This core principle of VE helps the engineer to identify solutions that quickly get to the root of the problem. It leads the engineer down a problem-solving path that identifies what needs to be done and arms him with factual data, instead of employing only what they want or think they should do.

DEFINITIONS OF VALUE ENGINEERING AND VALUE ANALYSIS

We engineers are very analytical in nature and have a fondness for mathematical equations. Before we go any further, we must define what value is and how it relates to VE and VA techniques:

$$Value = \frac{Function}{Cost}$$

where

Function all those things that the product, process, or procedure must do to make it work and sell.

Cost the expenditure of a resource such as time, money, people, energy, or material.

Thus, the primary objective of VE/VA is to improve value by one of the following ways:

- ❖ Reduce costs while increasing function.
- ❖ Reduce costs while maintaining function.
- ❖ Increase function while maintaining cost.
- ❖ Increase function while increasing cost by a lesser amount.
- ❖ Decrease function (in an over-specification situation) while decreasing costs by a large amount.

Meeting these objectives will allow us to reach our goal to obtain maximum function; performance; and prestige value of a part, product, or service per unit cost. Once this goal is achieved, there are many other opportunities in the engineering and manufacturing communities to increase a product's value. This is because the process by which the part is created is comprehensively understood. Some of the more prevalent causes for high-cost items—such as poor quality, high labor costs, excessive cycle time, use of premium materials, and so on—can now be investigated once the systematic approach of VE and VA has been executed. VE and VA are additional tools in an engineer's toolbox that allow us to efficiently and effectively solve problems and create value for our companies.

HOW IS VALUE ENGINEERING/VALUE ANALYSIS USED?

The concept of VE is not complete unless one employs VA. VA is a systematic approach that reduces a cost problem into seven phases:

1. ***Identification.*** The recognition of cost-saving opportunities, using the knowledge gained from both education and experience that helps to answer the question, "What does the customer want?" Here, the customer could be internal or external and is identified as the end user of the part, product, or service. You can have the best product in the world, but if the customer is unhappy with it, no one will buy it!
2. ***Information.*** Effective VA requires specific, quantitative, and qualitative information. During this phase, one must ask oneself, "How reliable is my data?" Time dependent data can be extremely useful here, and the organization of such data using spreadsheets and tables is key. The new engineer, who is early in his career, should remember to try to obtain information and develop feasible solutions before unnecessarily approaching others. Extract facts and data without irritating the sources of information that must remain valuable for future use. In other words, don't annoy the

senior engineers in your company with seemingly unintelligent questions, as they are the ones who will be vital to furthering your own knowledge and experience. You will know when you have frustrated them or wasted their time with a silly question or request, as they will likely make themselves unavailable when asked future questions or they will bluntly tell you so.

3. ***Speculation.*** Speculation is the creative phase in finding the solution to a cost problem. Benchmarking is often useful during this phase, as the individual or project team is able to discern how others do things. However, benchmarking alone will only allow you to be as good as the others you benchmark against. The successful engineer often takes what is learned by benchmarking, speculates on how this process can be improved, and implements an improved process.
4. ***Ideas.*** Additionally, brainstorming is essential here. During any brainstorming session, be careful to actively delay criticism of any new idea, and be certain that no new ideas are overlooked. The idea that seems to be the most outlandish at first glance is the one that is often implemented.
5. ***Evaluation.*** During the evaluation phase, the list of ideas that were advanced by speculation is shortened to the best two or three possibilities. Here, the list could be shortened by examining all qualitative or quantitative information.
6. ***Planning and Execution.*** It is important to have an effective plan in place that is organized and easy to follow. An effective plan keeps a project focused and on time. When the plan is put into action, one must be diligent in documenting how the plan was executed for the understanding of others.
7. ***Recording and Follow-Up.*** Key results or metrics should be recorded for an extended period of time, even after the project is complete. This phase helps to ensure that the project is meeting predefined goals and is on time. Even the problems or issues faced during the project should be recorded. Once recorded, display the results prominently so all team members can observe the progress of the project. If the results were positive, great! If negative, record them and let the team members know the problems are still being attacked and that further action is being taken.

STRENGTHS OF VALUE ENGINEERING CONCEPTS

One of the great things about VE is that it is not just limited to engineers. Effective execution of VE techniques involves all aspects of an organization, including manufacturing, quality control, purchasing, marketing, and human resources. When implemented appropriately, VE has resulted in:

- Significant cost savings
- Reduced lead times
- Better quality and performance
- Reduced product weight and size
- Shorter manufacturing cycle times

These types of results happen because often the entire company, not only the Engineering Department, is working in lock step to attack a particular problem. Because VE/VA techniques require that each step in the product life cycle be broken down and investigated, the process is more thoroughly understood by all; non-value added activities emerge more quickly. VE is a simple, yet powerful tool in reducing costs while improving product function and performance.

VALUE ENGINEERING WORKS!

If I were to tell you that VE was a great tool, but I never used it, would you feel confident that it actually works? Would you take my word at face value, without knowing me, my background, or experience? I know the answer to this is no, because you are an engineer. I've never met an engineer who would try something unproven "just because it sounds good." Engineers are very methodical, analytical people. I am an engineer, and I've lived this methodology. I know that when implemented correctly, it works, and I want to share my experience with you.

MY INTERNSHIP

When I first found out that for my summer internship I was going to work for a food company and deal with yogurt, I was less than thrilled. While all of my friends announced they were working for "sexy" high-tech companies in the defense or automotive industries, here I was, working for a food manufacturer. I was the first to admit that my internship wasn't the coolest place to work, and I was afraid of the jabs and jeering in store for me once my friends saw how silly I looked in the hairnet I was required to wear at the plant. All these things aside, I was happy for the opportunity, and even though this opportunity didn't seem to be the dream internship I was hoping for at first, I decided that I would make it into a great learning experience. I pledged to create value for the company that took a chance on hiring an inexperienced engineering student such as me.

To be successful at my new summer job, I knew that I needed to create some objectives. I decided that no matter what, I was going to make an impact on my employer so that maybe, just maybe, my internship could turn into a potential job lead. I decided to always be positive in spirit and attitude and to smile, as I knew a smile can go a long way. I wanted to be proactive, create value, and demonstrate the abilities that I knew employers were looking for, such as leadership, interpersonal skills, communication, and technical skills. While my internship didn't appear to be a dream opportunity at first, maybe I could turn it into one. I was determined to do that.

My internship assignment was to increase production line efficiency. This plant was the manufacturer of food products for various customers nationwide. There were four filling, packaging, and material handling lines in operation 24 hours a day, 7 days a week. It was vitally important that the equipment run efficiently and productively. Machinery downtime meant lost money. Poor quality or lost product meant lost money. These two things together not only meant lost money, but unhappy customers. I soon realized that my assignment was an important one, and I had the ability to make a direct impact on the bottom line of the company as well as on the overall satisfaction of the customers. Maybe this internship wasn't so bad after all.

Over My Head? Maybe Not!

My first day on the job consisted of a plant tour. I was told that the machinery was operating at 65% of its maximum capability and that there was excessive downtime of the machinery and excessive product waste. I was also told that the current design of some aspects of the machinery was insufficient, and all of these problems resulted in the company's inability to meet their production schedule and keep their customers happy. This seemed like an enormous task—was I in over my head? Did I have the experience and knowledge to be successful? I only had a few months and limited resources. What tools did I have at my disposal that would allow me to fix these problems and showcase my engineering skills? Ah, yes—I'm armed with the concepts of Value Engineering. When I thought of this, I breathed a great sigh of relief. Keep it simple. Use my common sense. My engineering education and career institute course prepared me well. I can do this.

I did not have much time to waste. I used VA as my solution method to this problem and redefined the stages of VA for this project in a more meaningful way.

1. *Identification and Information.* I knew I had to observe the operation closely and pay attention to detail. I'd conduct information interviews with employees and production line workers to find out why these problems were occurring. I'd research past historical data, and gather current production data, performance metrics, and any other relevant information that may help me determine the root causes of the problems.
2. *Speculation.* I knew I had to define and sharpen the focus of the problem. I'd have to brainstorm for possible solutions. Listing the goals of the project was essential:
   a. What do I want to accomplish?
   b. Are my goals feasible?
   c. What are my constraints—time, money, other resources?
3. *Evaluation.* All possible solutions must be investigated. Relationships between the data must be examined and trends analysis must be used. In coming up with my final solution plan, all information must be considered.
4. *Planning and Execution.* To ensure that my solution met my predefined goals, I had to plan, plan, plan. I had to plan around available resources, especially time—my internship was only for a period of three months. I had to execute my plan, to effectively put my plan into action. This meant I had to break down large tasks into smaller subtasks to make them more manageable. Also, everything I did, the things that worked and the things that didn't, had to be documented.

5. *Recording and Follow-Up.* I'd use recorded results as a measure and display for all employees to see. I'd organize and publish all of my documentation and make sure there were open lines of communication to me so that if anyone had recommendations, they would be considered.

Big Results through Value Engineering!

I painstakingly implemented VE/VA techniques for all three months of my internship. My solution plan included the redesign of some aspects of the production line, and by effectively putting it into action, I was able to achieve results that I didn't think were possible.

- Production line efficiency was improved to 94%.
- Machinery downtime decreased by 80%.
- Savings in lost product improved from 3,500 pieces per production run to 275.
- Resulting annual savings to the company was $60,000.
- The company's ability to meet their production schedule increased.
- Employee morale improved with improved work conditions and a better system.

The Final Result

My final recommendation was to invest in a new machine that would increase production capacity, decrease wasted product even further, and pay for itself after six months. I felt a high sense of achievement when I learned a few months after my internship experience that the company did follow my recommendation to buy the $110,000 machine, and it did pay for itself in six months. That alone represented a recurring annual cost saving of a quarter of a million dollars for my employer. Not bad from a student! This was, in fact, the dream internship I hoped for.

UNDERGRADUATE ENGINEERING CAREER DEVELOPMENT

How undergraduate engineering career development helped me can help you! Let me endorse an engineering career development program for you, an individual student with this book, or the school of engineering at any university. For example, my undergraduate engineering career development program was in many ways a life-changing event for me. The classroom and internship experience I gained and the successes I achieved during this time confirmed for me that I had what it takes to join the workforce and that I was "ready for takeoff." I gained not only the valuable hands-on engineering experience that young engineers so desperately need to jump-start their careers, but I also learned the soft skills that employers require today so that new hires can make an immediate impact from the first day on the job. In fact, VE is one of the subjects I learned in the course. The program gave me confidence in my abilities and allowed me to talk about the tangible accomplishments on my internship during future employment interviews. Now, after working in industry for over 12 years since graduation, I can truly say that my participation in an engineering career development program started me on the right path in my career. The program prepared me for numerous challenges, diverse assignments, and difficult management situations that I faced since graduation. I recommend an engineering career development type of course for any engineering school interested in preparing students for immediate professional success. Ultimately, I urge you to take personal responsibility for your own career development; you're the one who is in charge of your future career!

Chapter Summary: Takeoff Tips

- ❖ Value Engineering evaluates each step to manufacture a product that conforms to its intended functions and has the lowest possible cost.
  - Function: those things needed to make the product work and sell.
  - Cost: resource expenses such as time, money, people, energy, and material.
  - The primary objective of VE/VA is to improve value by optimizing cost and function.
- ❖ Value Analysis reduces a cost problem into seven phases:
  1. Identification: the recognition of cost-saving opportunities
  2. Information: specific quantitative and qualitative data
  3. Speculation: the creative phase to solve a cost problem
  4. Ideas: brainstorming
  5. Evaluation: consider the best two or three possibilities
  6. Planning and Execution: a plan that is organized and managed
  7. Record and Follow-Up: to ensure the project is meeting specified goals
- ❖ Big results through value engineering:
  - Rick Licursi achieved them on an internship after his junior year.
  - You can too.

Exercises

1. Define Value Engineering.
2. Define Value Analysis.
3. Describe, with an example, the seven phases of Value Analysis.
4. How did Rick Licursi achieve big results on his internship through Value Engineering?

CHAPTER 14

Quality Engineering

Quality Engineering, including Total Quality Management (TQM), Six Sigma, and the International Organization for Standardization (ISO) are very important throughout industry and government. Therefore, as you step into a job, and even an interview, it is advantageous for you to have some understanding of these quality systems. Once again, knowledge of these subjects will put you ahead of the crowd to compete for an offer and bring enhanced value to your prospective employer.

W. EDWARDS DEMING

As background, W. Edwards Deming is regarded as the father of TQM. Born in Iowa in 1900, Deming obtained a BS degree in Electrical Engineering and MS and PhD degrees in Mathematics and Physics. Dr. Deming's specialty was statistical process control (SPC).

During World War II, Dr. Deming taught SPC to U.S. companies to greatly improve their precision in manufacturing processes. The result was much better U.S. wartime products. Unfortunately, however, after the war Deming was not successful in convincing U.S. industries, such as automakers, to adopt SPC practices. Why was industry indifferent to Deming, who had proven SPC success during the war? Answer: complacency and huge, worldwide demand for American products.

With Europe and the Far East devastated following World War II, the United States produced more than half the gross world product; products "Made in USA" were even more prevalent than products "Made in China" are today. Comfortable with that success, U.S. industry did not feel the need to adopt Dr. Deming's different point of view as to how companies should be run or the importance of TQM. With plenty of customers and profit, there was no need to change. U.S. industry had an invincible mind-set and dismissed Dr. Deming's ideas as unimportant.

On the other hand, Japan, having been totally defeated in World War II, was much more in need of Dr. Deming's expertise to rebuild their industry and economy. During the period immediately following the war, goods made in Japan were regarded as low-quality junk. Then, in 1950 W. Edwards Deming stepped in. Big change and Japanese success ahead! Whereas U.S. industry brushed him aside, the Japanese embraced him as a hero with great ideas on how to produce quality products. Beginning in 1950, Deming trained hundreds of Japanese technicians, engineers, and managers in SPC and TQM. The result was that a totally defeated country quickly reached unprecedented levels of quality through the application of Deming's techniques. For example, Japan took manufacturing measurement tolerances that had been regarded as "best ever," and substantially improved standardized precision. One inch of variation tolerance became 1/16 of an inch. The world, and especially the United States, was to be stunned by this incredible leap forward by Japan.

JIM MELTON

Now, you will receive quality engineering information from Mr. Jim Melton, formerly Worldwide Manager of Quality Assurance for Motorola, Corp. and now Plant Manager of Continental Automotive, Inc. He received his BS Degree in Electrical Engineering from the University of Kentucky and his MBA from the University of Michigan. Mr. Melton is an expert in TQM and Six Sigma. He works with these concepts daily and has done so around the world. As an expert, Mr. Melton will explain the essentials of TQM and Six Sigma. This will be of great value to you as you enter the workplace.

WHY IS QUALITY IMPORTANT?

An answer to this is clear in the story of Dr. W. Edwards Deming. Countries and businesses that make reasonably priced and high-quality products are more likely to succeed than those who don't. In addition, when an individual, company, or country gets too complacent, they can be in for an unfortunate surprise. We all need to stay sharp. Always in shape. Always improving. Jim Melton gives us the following illustration of this.

An example is the U.S. auto industry. In the 1950s and 1960s General Motors Corporation (GM) had approximately 65% of the U.S. market share of automobiles. A common phrase at the time was "what's good for General Motors is good for America." GM was the premier blue-chip corporation. In fact, GM was so large, powerful, and successful that some people felt that it was a monopoly and should be broken up. Then what happened?

- ❖ The 1970s oil crisis presented an opportunity for more fuel-efficient vehicles, specifically Japanese, to gain in popularity.
- ❖ Superior quality and reliability of Japanese automobiles were recognized as offering a better value to the consumer.

- 30 years later, U.S. automakers are still trying to catch up. Toyota is now the largest automaker in the world; meanwhile GM had to face bankruptcy proceedings during 2009.

Mr. Jim Melton will now explain TQM.

TQM AND DR. DEMING'S QUALITY PHILOSOPHIES

- ***Quality Is a Management Responsibility.*** An organization that adopts TQM does so at the initiation and leadership of top management. Companies that lack top management commitment to it will fail in their attempt to install a TQM system. So, starting at the top, TQM becomes a system and culture that extends throughout the organization. Everyone is on board, committed, and personally involved.
- ***Customer Focus.*** A TQM organization believes that it will only be successful if customers are given top priority. Customers are external as well as internal. The importance of external customer focus is obvious. Within a TQM organization, some functions are customers and others are suppliers. For example, design engineering would be a supplier to their customer, manufacturing. Attention would be given internally as if manufacturing were an external customer. This internal customer focus is especially new, and sometimes challenging, to an organization transitioning to TQM.
- ***Continuous Improvement.*** Japanese call it "kaizen" and it is integral to a TQM organization. The process being used today must be examined and improved upon. All activities within the organization are subject to measurement and subsequent improvement. Emphasis is on continuous improvement—getting better and better.
- ***Reduce Process Variation.*** Processes tend to have normal variation in performance. If you took a dime and dropped it repeatedly, it would land in different places each time as it bounced or rolled. Dr. Deming thought that this kind of process variation was the root of all evil in obtaining quality. Normal process variation must be tightened.
- ***Statistical Process Control (SPC).*** Process variation can be understood and monitored through statistical process control. This means avoidance of tampering with processes in an attempt to control them when the process is in its normal range of variation. For example, in a manufacturing process that includes soldering, there could be an unacceptable variation among workers or equipment as to the amount of solder applied to a joint. In such a case, this is a problem that must be solved through SPC.

  If you have an excellent product that is popular with customers, they will expect consistency. Lacking consistent quality, a customer would go to the competition. So, if you have an excellent muffin recipe and want to be in the muffin business, you have to have the same recipe, ingredients, and process—including oven temperature, bowls, and so on. With that outstanding recipe and process consistency, your muffin will taste the same in Boston as it does in Waco, and you have taken the first step to dominate a niche in the muffin market!

- ***Understand the Process Capability.*** Products must be designed to accept the process capability. Otherwise, the process capability must be improved.

    Dr. Deming was very strong in his commitment to control and understand processes. His work in applying SPC focused on eliminating process variation. This brings us to the concept of Six Sigma, which Jim Melton will explain.

SIX SIGMA

- Sigma, the 18th letter of the Greek alphabet, is the value of the process standard deviation for a given characteristic in a normal spread of data. This quantity is used to quantify the spread (around a mean) of some process or product characteristic.
- Sigma level is the business metric used to indicate the performance of a process to some specification.
- It uses the number of standard deviations we can fit between the mean and the nearest specification.
- It measures the number of defects per opportunity produced by a process. The sigma level is used to account for complexity that allows for the comparison of dissimilar goods and services.
- The goal is to produce goods and services at a Six Sigma level. As an organization moves toward its Six Sigma quality, they will:
    - Eliminate defects
    - Reduce production and development costs
    - Reduce cycle times and inventory levels
    - Increase profit margin and improve customer satisfaction
    - Achieve the highest sigma level possible

Why Is 99% Good Not Good Enough?

Offhand, 99% good would seem to be fine. In reality, 99% good would mean:

- 200,000 wrong drug prescriptions each year
- No electricity for seven hours each month
- Unsafe drinking water for 15 minutes each day
- 20,000 lost articles of mail per hour
- 5,000 incorrect surgical operations per week
- Two landings that are long or short at most major airports each day

So is 99% good, good enough? This evidence shows we need to do better than 99% good. Jim Melton will now explain the sigma levels that improve upon 99% good.

Sigma Levels

- *Historical Standard: 3 Sigma Capability.* 3 sigma was accepted for many years as the highest standard of excellence. It was a big improvement over previous levels of quality, representing 93.319% accuracy in a process. Not good enough!
- *Current Standard: 4 Sigma Capability.* 4 sigma level is currently regarded as a high level of quality. It substantially improves on 3 sigma in that it represents 99.379% process accuracy. Might seem to be as good as it gets, but wait. What's next?
- *New Standard: 6 Sigma Capability.* This new standard of quality excellence represents another huge leap forward in TQM. Invented by Motorola, companies such as General Electric, many other corporations, and government organizations now operate at this high level. Incredibly, Six Sigma means that your process is 99.99966% accurate. This seems impossible, but it's being done every day in the most progressive, quality organizations. It means that only 3 parts per million (ppm) are defective. This is great. How do we get there? How do we do it?

Six Sigma Success Strategy

Here's the method, which Jim Melton will explain.

- *Characterization.* Characterization means distinguishing or identifying a part of a process that needs to be studied and improved upon. Characterization is the first general stage of Six Sigma problem-solving. Specifically, you will:
  - *Define the Project Scope.* What problem are you trying to solve? A real-life example is an electrically conductive adhesive with an unacceptable range of viscosity in the process of manufacturing a screen-printing product.
  - *Measure.* The measurement process means we select desirable output characteristics (Y's) to be measured. We then assess performance specifications and validate measurement systems. Finally, we establish an initial capability of measurement; that is, we answer the question: "What is the desired measurement process?"

    In our real-life example, we need to establish the inputs and outputs to be measured. Moreover, we need to establish whether the measurement system is accurate and repeatable. It turns out that viscosity measurement is not repeatable. Therefore, there is a need to change measurement equipment and technique to create stable readings. Now, after measurement modifications, we have accomplished that.
  - *Analyze.* During this final stage of characterization, we define performance objectives, and then we document potential problems (X's). We will then analyze sources of variability to be reduced.

    In our real-life example, we need to know the key parameters to control. Usually this is determined by design of experiment (DOE) methodology. What is the recipe to get the right viscosity each time? What is the tolerance range that is acceptable? In this case, we need to determine what in the adhesive has the greatest impact on viscosity. There are five variables, but we assess one at a time while holding the other four constant. What changed? Which part has variation? The answer is silver. Then we need to determine what characteristic of the silver has the greatest impact. The answer is the size of silver particles.

- ***Characterization Summary.*** Now we have completed our first part, characterization steps, in Six Sigma problem-solving. We have defined our problem, measured the output characteristics, and analyzed our problem, including the source of variability. In this example case, analysis proved that the problem was the variable size of silver particles. Having defined the part of the process that needs improvement, we will go to the solution stage—optimization.

❖ ***Optimization.*** During the characterization steps, we have defined the problem, measured, and defined the key process variation problem through analysis. Optimization is the next general strategy stage of Six Sigma. Optimization means we have to improve and control, as follows.

- ***Improve.*** This means we need to screen potential causes and identify appropriate operating conditions.

  In our real-life example, we need to implement the changes that the data has revealed. This could involve equipment, material, or product design. After studying problem causes and options, we reach the improvement conclusion that we need to use batches of silver that have the proper grain size.

- ***Control.*** When we control, we need to determine process capability. Then we implement process controls. Finally, we document what we have learned. Documentation is key: we check to ensure that we have the correct output, and then we write it down. Documentation enables the transfer of knowledge to others throughout the process, now and in the future. If we don't document, we have "word-of-mouth folklore" that destroys the precision solution we have accomplished and puts us back to the original problems of process variation.

  In our real-life example, we will use our improvement conclusion. Specifically, we will confirm the proper grain size in the silver prior to mixing the adhesive and verify that viscosity prior to shipment. These conclusions will be documented through changes to specifications. Moreover, the controls will be monitored through SPC—this means we error proof the process. In this, and any other SPC error-proofed application, equipment is programmed to stop the process when a variation (error) is identified.

We have now arrived at Six Sigma level of quality. We have a process that is 99.99966% accurate. Not long ago, no one would have believed this to be possible. Now, increasingly, it is a global standard. That leads us to our next Quality Engineering subject, the International Organization for Standardization.

INTERNATIONAL ORGANIZATION FOR STANDARDIZATION

In his national best-selling book, *The World Is Flat*, Thomas Friedman describes the explosion of business globalization in the twenty-first century. In summary, Friedman explains that the United States is on a business and technical level playing field with India, China, and other countries. That means that to be competitive, TQM and Six Sigma are concepts that are being used throughout the world. That makes the International Organization for Standardization (ISO) more important than ever, and therefore something you need to know about.

Definition

Founded in 1947, ISO is the world's largest quality standards-setting organization. With globalization, its growth and importance have increased significantly during the past 10 years. Now, ISO is a network of standards organizations of 158 countries. There is one national coordinating member per country with a central headquarters in Geneva, Switzerland. ISO is derived from the Greek word *isos* that means "equal." In all countries, in all languages, the organization's name is ISO.

ISO 9000 is a grouping of quality management standard systems that especially relates to engineering, as follows.

- **ISO 9001** is a quality assurance standards system for development, design, production, installation, and service.
- **ISO 9002** is a quality assurance standard system for production and installation.
- **ISO 9003** is a quality assurance standard system for inspection and testing.

ISO 14000 is also very relevant to engineering; this grouping of quality management standards addresses environmental management. The goal of ISO 14000 is to ensure that organizations minimize environmental damage and work to improve their environmental impact.

As a practicing engineer, you will be involved with helping your organization meet and maintain ISO accreditation. This accreditation is established by an independent certification group. The organization you are likely to work for is examined and registered as meeting one or more of the ISO standard systems above. Organizations use this certification as evidence that they can deliver quality service to customers anywhere in the world.

Now, thanks to Mr. Melton, you know about quality engineering, including TQM, Six Sigma, and ISO. You are likely to be using these concepts on a daily basis. Throughout your career, you will ensure that your individual and team engineering work are done at the highest standard of quality. That is the only way that the United States will remain competitive and allow us to enjoy our high standard of living now and in the future.

Chapter Summary: Takeoff Tips

- ❖ Countries and businesses that make reasonably priced and high-quality products are more likely to succeed than those that don't.
- ❖ Total Quality Management (TQM) and Dr. Deming's quality philosophies:
  - Quality is a management responsibility.
  - Customer focus—customers are given top priority.
  - Continuous improvement.
  - Reduce process variation—the core challenge in obtaining quality.
  - Statistical process control (SPC)—enabling monitoring of process variation.
  - Understand the process capability.
- ❖ Sigma: the value of the process standard deviation for a given characteristic
- ❖ Six Sigma: the new standard of quality—99.99966% accurate
- ❖ Six Sigma success strategy:
  - Characterization: that is, define, measure, and analyze
  - Optimization: that is, improve and control
- ❖ ISO 9000: international quality management standards for engineering

Exercises

1. Describe the significance of Dr. W. Edwards Deming and SPC.
2. Why is quality important?
3. State Dr. Deming's philosophies for total quality management.
4. Describe the Six Sigma success strategy.
5. What is ISO? Why is it important to an engineer?

CHAPTER 15

Lean Enterprise

What is Lean? Lean means producing goods and services in the most effective way possible. It centers on removing wasteful steps and creating the smoothest, most efficient flow process while focusing on the needs of the customer. Lean enterprise is big. It blends with Project Management, Value Engineering, and Quality Engineering that were covered in the last three chapters. Specifically, Lean is very important for you as an engineer to know as you begin your employment. It is important whether you are engaged in manufacturing or any other area of enterprise.

To explain Lean and its importance to you as an engineering student, we call on the wisdom of Dr. John Voit. Dr. Voit is the Lean Implementation Manager of New Product Development for an automotive supplier. Dr. Voit holds a PhD in Industrial and Systems Engineering and is an expert on Lean. He will, in this chapter, explain Lean applications for the benefit of your employer and yourself.

BRIEF HISTORY OF MANUFACTURING

- *Medieval and Pre-Modern Craftsmen.* If you were wealthy and wanted a product or work of art, a craftsman could provide it to you. The quality of work among craftsmen varied; some were extremely proficient and some were not very good. If you wanted a portrait painting of your spouse, you might be happy with the result or quite disappointed. In short, there was great variation in quality, cost was high, and it took much time to create a product.
- *Nineteenth Century: Interchangeable Parts.* Eli Whitney is best known as having invented the cotton gin. Later, he focused on manufacturing firearms with interchangeable parts. This was of great military importance because if a part of a weapon failed, a new part could replace it; the weapon would not have to be thrown away. The concept

of interchangeable parts became significant with developments by Henry Ford in the early twentieth century.

- ❖ ***Mass Production: Henry Ford.*** In 1913, Henry Ford combined interchangeable parts with the standard work practices of Frederick Taylor to create the automobile manufacturing assembly line. Ford called the process "flow production." The result was a dramatic reduction in production time and major improvements in cost and quality. One problem with mass production, however, was a lack of variety. Of the Model T Ford, Henry Ford said, "You can have any color you want, as long as it's black." Why did he choose black? It was technically logical: black paint dried faster and that saved time. Eventually consumers wanted variety, and other auto manufacturers responded to that desire.
- ❖ ***The Toyota Production System (TPS).*** Kiichiro Toyoda and Taiichi Ohno of Toyota Motor Company closely studied Henry Ford's success and improved upon it. This and the contributions of W. Edwards Deming, described in Chapter 14, resulted in TPS. The TPS approach used the efficient production system developed by Henry Ford and took it further by introducing a wider variety of options for the consumer. TPS also introduced lower, more precise, production volumes and a reduced number of defects. TPS and Lean were linked in a best-selling book, *The Machine that Changed the World: The Story of Lean Production—Toyota's Secret Weapon in the Global Car Wars that Is Revolutionizing World Industry* by James P. Womack, Daniel T. Jones, and Daniel Roos. This book was the first to describe the much-improved concepts of Lean and how they can benefit both companies and their customers.

WHAT ARE THE SPECIFICS OF LEAN AND HOW AM I GOING TO USE THEM?

The essentials of Lean that you will want to apply on the job or at home are the seven forms of waste and five Lean principles. Dr. Voit will explain them, in that order.

Seven Forms of Waste

Examine your personal and employer's processes to find these seven forms of waste.

1. *Correction*—mistakes. Examples are quality defects resulting in product recalls or patching an item that was done incorrectly.
2. *Overproduction*—production exceeds demand. Examples are printing too many copies at your office and cooking too much for dinner—more food than can be eaten that is then discarded.
3. *Material Movement*—non-value adding activity of moving something. An example is laundry. Increasingly, home washers and dryers are located on first or second floors rather than basements.

4. ***Motion***—people or equipment moving more than is required. An example is driving to a store when you can purchase a product online instead.
5. ***Waiting***—waiting for the next production outcome. An example is standing in line at the Department of Motor Vehicles.
6. ***Inventory***—all material not being used. An example is storing 32 rolls of toilet paper.
7. ***Processing***—things done that are unnecessary. An example is a restaurant waiter writing down your order on a pad rather than use a computer. Here, however, is an example of prioritizing customer preference (that which people are willing to pay for) over pure efficiency.

   It is unlikely that you can walk into any operation and totally eliminate waste. Understanding the seven forms of waste, however, will allow you to identify areas of waste that can be removed or significantly reduced. With this knowledge, Dr. Voit will go on to explain five principles that are the foundation of Lean enterprise, adapted from *Lean Thinking,* by James P. Womack and Daniel T. Jones.

Five Lean Principles

1. ***Specify Value in the Eyes of the Customer.*** Who is a customer? People with cash. Value added (VA) is something that meets the customers' needs and for which they are willing to pay. Non-value added (NVA) activity is everything else.
2. ***Identify the Value Stream and Eliminate Waste.*** Identifying the value stream requires that all steps, both VA and NVA, be identified. Then, NVA steps can be eliminated. Examples include any of the seven forms of waste, including material movement, motion, waiting, and so on. Sometimes we are so focused on cost efficiency that we do not consider the wants and needs of the customer. This principle tells us not to eliminate something that the customer values. It may not be the most efficient, but if the customer wants it and is willing to pay for it, it has value. For more on this, a good book is *Learning to See: Value Stream Mapping to Create Value and Eliminate MUDA* by Mike Rother and John Shook.
3. ***Make Value Flow at the Pull of the Customer.*** Let's describe flow and pull.
   - ***Flow.*** Flow means creating a value stream so the product goes from one VA activity to the next most quickly. NVA activities interrupt flow. The flow of manufacturing a product is based on customer orders. An example is a computer company assembling computers only after orders with specifications are submitted, not before.
   - ***Pull.*** Pull means creating a value stream that is responsive to customer demand. This is unlike a push system that continues to generate a product when there is no clear linkage to customer orders. This may lead to overstock on some products and shortages on those that the customer really wants.

4. ***Involve and Empower the Employees.*** Respect employees and solicit their ideas. They work with the processes every day and can be best able to suggest ideas that minimize waste while improving value and efficiency.
5. ***Continuous Pursuit of Perfection.*** This means continuous pursuit of a goal of eliminating waste and producing value at the pull of the customer. It also can include adding automation and new technology when it is justified. Realize this:
   - Pursuit of perfection is an incremental journey: step by step, year after year.
   - Even some of the leanest value streams still have more than 50% waste.
   - Some waste is necessary. Inventory is waste, because we know of no better way to do it today, but it is still waste.

Dr. Voit has just summarized Lean enterprise, including the seven forms of waste and five Lean principles. While Lean was first introduced for more efficient manufacturing processes, it can apply to many other areas. That includes:

- ❖ Product development
- ❖ Order taking and scheduling
- ❖ Logistics
- ❖ Administrative systems
- ❖ Human resources
- ❖ Everything

So, make the most of your knowledge of Lean and be a more valuable, successful employee. As indicated, Lean can help us be more efficient in our personal lives too.

Our lives, personal and professional, include eliminating waste and improving things. What's out there and how can we make it better?

Chapter Summary: Takeoff Tips

- ❖ Evolution of Lean
  - Pre-modern craftsman: great variation in quality, high cost, much time
  - Nineteenth century: interchangeable parts and Eli Whitney
  - Twentieth century: mass production and Henry Ford
  - The Toyota production system (TPS)
- ❖ Seven Forms of Waste
  1. Correction
  2. Overproduction
  3. Material movement
  4. Motion

5. Waiting
6. Inventory
7. Processing

❖ Five Lean Principles
1. Specify value in the eyes of the customer
2. Identify the value stream and eliminate waste
3. Make value flow at the pull of the customer
4. Involve and empower the employees
5. Continuous pursuit of perfection

Exercises

1. What is Lean? Why is it important?
2. Describe the Toyota production system (TPS).
3. What are the seven forms of waste?
4. What are the five Lean principles?

CHAPTER 16

Engineering Professionalism, Ethics, and Responsibility

Knowledge of engineering professionalism is vitally important for your career. Let's define profession: according to Webster's dictionary, it is a vocation "requiring specialized knowledge and often long and intensive academic preparation." Since you are experiencing "long and intensive academic preparation," you know that you're headed toward a profession. And it's an extremely respected profession at that. Engineers consistently rank among the 10 most respected professions of all. With that respect comes your requirement of maintaining a professional level of competence, ethics, and responsibility.

In this chapter, Ms. Maria Lehman will explain in detail what it means to be a true engineering professional along with ethical responsibility and liability that go together.

Ms. Maria Lehman is Vice President and Business Director of Transportation Systems for Bergman Associates. Previous positions include Chief Operating Officer for the Chazen Companies and Vice President for URS Corporation. She has significant experience in the public sector as well, having been Commissioner of Public Works of Erie County, New York, with a $100 million capital budget. She has a BS degree in Civil Engineering, is a registered Professional Engineer, and has been a board member and National Vice President of the American Society of Civil Engineers (ASCE). Ms. Lehman is an expert on the subject of engineering professionalism and has valuable perspectives that will help you in your career.

ATTRIBUTES OF THE ENGINEERING PROFESSION

- Engineering is indispensable. Our standard of living depends on engineers. The public needs to have a better understanding of the importance of engineering in their daily lives.

- Engineering satisfies a broad range of needs through private enterprise as well as government service. Very few public leaders, including members of Congress, have engineering degrees. Engineers need to be vocal citizens—catalysts for change in public policy as needed. As an engineer, your leadership and communication skills will be vital as we progress into an ever more technical world.
- Engineering requires applied knowledge as well as discretion and judgment. Currently, per *Science and Engineering Indicators 2008* published by the National Science Board, the United States graduates approximately 70,000 engineers annually at the bachelor's level. China, meanwhile, graduates over 400,000 entry-level engineers per year. Despite the numeric shortfall, the United States has a competitive edge that we need to keep. Specifically, graduating engineers in the United States tend to be more able to apply their knowledge beyond theory learned in the classroom. That's huge.
- An important objective is the promotion of engineering careers, knowledge, and ideals for the public good. An interesting estimate: it is reputed that in the United States there are nine lawyers for every one engineer; in Japan there are nine engineers for every one lawyer. Is that proportion good for the United States? One thing for sure: as a future engineer, the United States needs you!

WHAT IS AN ENGINEERING LICENSED PROFESSIONAL?

The Engineering Practice Act protects the health, safety, and welfare of the citizens of any given state. There are 51 jurisdictions, including each state and the District of Columbia. Having an engineering license means more than just meeting a state's minimum requirements. It means you have accepted both the technical and ethical obligations of the engineering profession.

What Is a Licensed Engineer?

The professional engineering license grants you the opportunity to perform engineering services for the public and take responsibility for your designs, reports, and professional opinions. With that, you have the privilege of applying your state-authorized "engineering seal" to your engineering work.

Licensing

Licensing is a collaboration of education, industry, and government. If you plan to obtain an engineering professional license, you should be aware of the following licensing roles played by education, industry, and government that will affect you.

- ***Education.*** You must have at least a four-year ABET-accredited degree to be a professional engineer (PE). ABET stands for the Accreditation

Board for Engineering and Technology. An ABET accrediting team visits engineering schools every six years to ensure that each of its engineering departments is up to standard. That includes standards of excellence in teaching engineering fundamentals, as well as broad-based application and professionalism skills that are being taught in this chapter and book. At the end of your education, you may pass a Fundamentals of Engineering (FE) exam sponsored by the National Council of Examiners for Engineering and Surveying (NCEES). This exam is offered in April and October each year. Students can take the exam before or after they graduate. After passing the exam, you will be known as an engineering intern (EI) or engineer in training (EIT).

- ❖ ***Industry.*** After receiving your ABET-accredited engineering degree, you need to have four years of progressive engineering experience. Note: there are many states that accept co-op experience as part of this four-year requirement; check with the American Society of Engineering Education (ASEE) for your individual circumstance. It is advisable to join an engineering professional association in your major or one related to your work. Overall, you need to obtain experience and develop the skill of communicating what you know. While gaining experience, you should get to know the requirements of the state where you plan to have licensure. At the end of your required employment experience, you will be required to take a Principles and Practice of Engineering exam.
- ❖ ***Government.*** After four years of acceptable experience as an engineer in training, you are qualified to take the PE exam. You will do this in one or more of the states of your choice. After taking and passing the PE exam, you will be a licensed professional engineer, with the requirement that you maintain continued professional competence. State governments administer licenses and have various requirements for continuous professional development.

Why Should I Become Licensed?

- ❖ ***Public Recognition.*** As a licensed engineer, you achieved an enhanced status in the eyes of the public that equates you with professionals licensed in other fields such as physicians, attorneys, and accountants.
- ❖ ***Private Practice.*** If you think you may want, at any time in the future, to be a consulting engineer, own an engineering firm, or be in a responsible position in the public sector, you must be licensed.
- ❖ ***Public Practice.*** Many government agencies require that higher-level engineering positions be filled only by licensed engineers.
- ❖ ***Changing Workplace.*** Today's careers are characterized by more mobility than ever. During your lifetime, you may work for 10–15 employers and have multiple career changes. This means learning to adapt to change and

possibly transitioning into a consulting relationship with an employer. Such a relationship may require an engineering license.

- ***Ethical Responsibility.*** State licensing boards have standards of ethical conduct that are legally binding. The recognition and enforcement of licensing board standards gives greater definition to the profession and enhances the public image of the engineer.

ETHICS—THE CORE OF THE ENGINEERING PROFESSION

- They describe the high standards of behavior for the profession.
- Ethics are the rules of the road; that is, what you should do on a daily basis.
- Ethical standards change with time, based on technological, cultural, and legal changes. The following are legal now, but:
  - Ten years ago it was unethical and illegal to electronically sign a document.
  - Fifteen years ago it was unethical and illegal to fax a contract.
  - Fifty years ago it was unethical to market engineering services.

There are many shades of gray, and in view of constant changes, you need to know ethical behavior from a number of perspectives.

Individual Ethics

Individual ethics includes your personal philosophy and moral standards. Your personal ethics require awareness and sound judgment. For example, in third world countries, bribes are commonplace in business. In the United States, bribery is illegal. If you are employed by a U.S. multinational company and caught issuing a bribe in a third world country, you can be prosecuted in the United States. This is just one example of the need for personal awareness, self-discipline, and high ethical standards of behavior.

Stewardship also falls under the heading of individual ethics. Now, more than ever, engineers are responsible for sustainable development. Energy and environmental concerns are linked, and they are vital.

The United States has been a leader in environmental cleanup. However, we are behind on alternative energy and "green design." During your engineering career, you may have the opportunity to make great contributions in these areas. Such contributions will be ethically rewarding as well as having the potential of career success by any definition.

Professional Ethics

Visit the National Society for Professional Engineers (NSPE) Web site (www.nspe.org) for a detailed explanation of the Code of Ethics for Engineers. In summary, the *Fundamental Canons* say that "Engineers, in fulfillment of their professional duties shall:

1. Hold paramount the safety, health, and welfare of the public.
2. Perform services only in areas of their competence.

3. Issue public statements only in an objective and truthful manner.
4. Act for each employer or client as faithful agents or trustees.
5. Avoid deceptive acts.
6. Conduct themselves honorably, responsibly, ethically, and lawfully so as to enhance the honor, reputation, and usefulness of the profession."

In addition to the NSPE code of ethics, there are individual discipline codes of ethics. Specifically, if you have a student chapter of ASCE, AIChE, ASME, IEEE, IEE, AIA, and so on, you can benefit by joining as a student member, networking and learning more about the specific code of ethics for your discipline. Student chapters have joint meetings with parent chapters where you can meet prospective employers.

When you reach the level of professional registration, there are state-mandated laws that you want to know as standards of practice for that state.

The bottom line on professional ethics is that you'll stay informed and use your professional judgment on an ongoing basis to ensure you are acting honorably.

Workplace Ethics

- ❖ *Employment.* Your employer will have policies and procedures that you need to follow. If your employer has an employee manual, you want to read it at the start to ensure you are operating within procedural and ethical guidelines. Rules vary among states and employers, but as a liability risk question, ask yourself, "Would a reasonable engineer do it this way?" In addition to the employee manual, your employer may have conflict of interest certifications, software usage policies, and intellectual property agreements—all of which you will want to realize and act accordingly. For example, a conflict of interest problem occurs when a person tries to hide an inappropriate business relationship. Ask questions as necessary and get assurance that you're doing the right thing. Conflicts of interest are not viewed as conflicts if you disclose them and act appropriately.
- ❖ *Contractual Work.* You may be working as an independent contractor, working for one or more clients. Each client may have their own code of ethics that you need to be aware of. It is very important to know that your client's code of ethics do not conflict with your own high standards of professional responsibility. There are also government-imposed codes of ethics and standards of care in specific situations that you need to know about.

LIABILITY

As an engineer, you are responsible for your work. This discussion of liability is not intended to scare you, but rather, to enlighten you and reduce the risk of liability problems due to ignorance of this subject. Liability is real and you need to know about it.

Some Areas of Liability

- ❖ A design or construction defect.
- ❖ Negligent architectural or engineering design.
- ❖ Negligent field survey or field investigation, such as doing insufficient soil borings before construction, resulting in costly damage.
- ❖ Coordination errors. For example, if drawings are not indexed, the contractor may do it the wrong way and the person not indexing would be liable.
- ❖ Negligence by contractor, construction manager, subcontractor, materials supplier, property manager, or owner. For example, a subcontractor may be focused on doing the work at minimal cost, and perhaps skimp on quality or safety matters. If it's your project, you need to be sure that the work meets professional standards.

Determining Responsibility

How is it proven that you are responsible for an engineering defect?

- ❖ By establishing that you knew about the standard of care and chose to fall below it, specifically by creating, causing, or contributing to a defective condition.
- ❖ Expert witnesses are generally used to establish fault. Presume that they may be a competitor, and want you to look bad.
- ❖ The remedy to avoid all this: know the engineering standards of care and make it your firm practice to uphold them.

Lawsuits

What can an engineer be sued for?

- ❖ Redesign costs—fixing a bad design.
- ❖ Cost of repairing defects—construction costs.
- ❖ Expert witness fees—paying for a potential competitor to prove that you made a mistake.
- ❖ Expenses in evaluating the cause of defects.
- ❖ Court and litigation costs.
- ❖ Punitive damages—if it's proven you did something wrong on purpose.
- ❖ Loss of use, loss of rents, or loss of market value: that is, loss of profit.
- ❖ Lawsuit costs.
- ❖ Professional malpractice insurance. Your employer should have it and you should be covered. If there is a claim, the insurance premium goes up, perhaps a lot.
- ❖ The amount of insurance deductible. With insurance costs rising, there is a trend toward higher deductibles.
- ❖ Time spent on defense is not billable time. That is an opportunity cost.
- ❖ Intangible costs: reputation—a widely published negative news story can do substantial damage to a firm's reputation, and thus revenue.

- Emotional impact—there is a negative emotional cost associated with a lawsuit. Consider this the "look in the mirror" syndrome; you need to like who you see. Consistently acting with high standards of professionalism will reduce the chance you have to experience a lawsuit.

Best Defense

What is the best defense to protect you from liability problems?

- A well-prepared design-professional contract, incorporating a favorable indemnity clause and a clause requiring owners to maintain their own insurance.
- A contract that clearly spells out what you are and are not doing for a fee. During the development of a contract, a lot of things are negotiated in or out of the contract. It is essential to document what's in the contract and what's out. Obviously, it is very important for you to read and fully understand the contract, especially what you are responsible for.
- Well-executed design and inspection. In short, do a good job.
- Thorough documentation. This can include e-mails, letters, meeting minutes, and telephone notes. When a decision is reached through discussion, protect yourself by writing a quick e-mail documenting the conclusion. Don't assume that because you said it, people will remember what you said.
- Independent technical reviews. Get an independent review before completing an important design project. Have a well-qualified person look at the proposed project and ask: Can I build this? Does it make sense?
- Pay attention to details and proper coordination in drawings and specifications.
- Avoid redundant information, especially if it is not electronically linked.
- Be sure that your words and sentences are clear, make sense, represent your true intentions, and are understood by you and your client.
- Do not rely on client or regulatory signoff as a defense. Your client or a regulatory authority might want to save money by skipping an important safety element in the design. If it smells, if it feels questionable, or if you are uncomfortable with what you are being asked to do, communicate this information to the appropriate people and, importantly, keep documentation of your concerns.
- Always have a contract, no matter how small or large the project, even if it is only a basic purchase order.
- As much as possible, use indemnity agreements (hold harmless agreements). These agreements exempt you from penalties or liability.
- Use standard agreements, such as from the Engineers Joint Contract Documents Committee (EJCDC) found on the Internet at www.acec.org/publications under the contracts/forms/guides link.
- Notify the insurance company immediately of a potential claim, or you might jeopardize your right for coverage.

Prepare Yourself for Engineering Ethical Responsibility

- Most engineering education focuses technically on "how to do things right."
- During employment, you will need to "do the right thing."
- Doing things right and doing the right thing are both very important.
- Communication: ask questions if you are not sure. Your employer or your professional organization may help you.
- Understand the rules of professional ethics and develop the ability to recognize an ethical question. That said, realize that ethical questions are not always easy to answer.
- Be willing to fight for what's right. That may be despite time and budget constraints, internal political pressure, and personal considerations.

Facing an Ethical Decision

Ms. Lehman quotes the former Vice President of Camp Dresser McKee, Mr. Charlie Parthum, PE. He is also a former National ASCE President. He presented these six ethical questions for any engineer:

1. Is it right?
2. Is it fair?
3. Who gets hurt?
4. Can it be made public?
5. Would you tell your child to do it?
6. How does it smell?

PREPARE YOURSELF

Maria Lehman's summary for engineering professionalism:

- Strive for continuous improvement.
- Continue your education: technical, business, or legal, as appropriate.
- Develop your soft skills, including communication and leadership.
- Embrace change.
- Nurture a teamwork philosophy.
- Learn to communicate honestly and effectively. You may need to tell your boss, client, or colleague if something's wrong. Tell them nicely. Most will appreciate it at the end.
- Share your knowledge outside your technical circle. Be aware of what's going on.

Be proud and happy to be an engineer. It is a great profession. You will be building and improving products or processes that solve problems, create opportunities, and enrich lives. Your positive contributions add up to why, in survey after survey, engineers are ranked among the most highly respected professions. Nice to have that respect while doing good things, making money, and having fun!

Knowledge is power, and acting on this knowledge will not only protect you, but will put you well ahead in your engineering career.

Chapter Summary: Takeoff Tips

- Attributes of the engineering profession:
  - Our standard of living depends on engineers.
  - Engineering requires applied knowledge plus discretion and judgment.
  - We must promote engineering careers and ethics for the public good.
- Licensing is a collaboration of education, industry, and government:
  - ***Education.*** You must have at least a four-year ABET-accredited degree.
  - ***Industry.*** You need four years of progressive engineering experience.
  - ***Government.*** Following your ABET-accredited engineering degree and four years experience as an engineer in training, you are qualified to take the Professional Engineer (PE) exam.
- Ethics is the core of the engineering profession:
  - ***Individual Ethics.*** Your moral standards, awareness, judgment, stewardship.
  - ***Professional Ethics.*** As explained in the Code of Ethics for Engineers.
  - ***Workplace Ethics.*** See your employee manual or client's code of ethics.
  - Be sure they do not conflict with your own high ethical standards.
  - There may also be government-imposed ethics and standards.
- ***Liability.*** As an engineer, you are responsible for your work.
- The best defense to protect you from liability problems:
  - A well-prepared design/professional contract.
  - Well-executed design and inspection.
  - Thorough documentation including e-mails, letters, meeting notes.
  - Get independent technical reviews before completing an important project.
- Maria Lehman's summary for engineering professionalism: See her summary on page 168. Do it. Be it.

Exercises

1. What is a licensed engineer?
2. What are the advantages for you in becoming licensed?
3. Why are ethics at the core of the engineering profession?
4. Define engineering liability.
5. What is the best defense to protect you from liability problems?
6. What are six questions to ask yourself for an ethical decision?

CHAPTER 17

Entrepreneurism

INTRODUCTION

So far, Part 2 of *Ready for Takeoff!* has focused on professional functions and being successfully employed in existing engineering organizations. Since the objective of this book is a total engineering career launch, you need to be introduced to a full range of engineering career options. Therefore, in this chapter we'll cover entrepreneurism and Chapter 18 will introduce you to academic careers and graduate school. Now, let's talk about a career as an entrepreneur.

WHAT IS AN ENTREPRENEUR?

An entrepreneur is a person who starts a business and is responsible for its outcome. Let us salute the entrepreneur. Every private enterprise, large or small, had to be founded by someone, whether it was Henry Ford, the Wright brothers, or Bill Gates. The United States has led and inspired the world with its ingenuity, entrepreneurial spirit, and free enterprise system. Immigrants coming to the United States have been especially awed by and successful with the opportunity to start their own businesses. Indeed, right now first-generation immigrants are reputed to be more likely to become self-made millionaires than those who were born in the United States. We need to let that sink in when we consider our affluent society that some people take for granted, including expected reward (entitlement) without their corresponding effort and calculated risk. Affluence and apathy can bring some surprises: 2,000 years ago "all roads led to Rome" and during the nineteenth century "the sun never set on the British Empire"—but let's get on to our subject of entrepreneurism today.

COULD YOU BE AN ENTREPRENEUR?

Today, engineering entrepreneurs generally have begun working somewhere where they get a bright idea and a passion to build it into their own business.

Engineering entrepreneurs have a passionate dream and the technical expertise, determination, and business savvy to turn the dream into reality. The most successful can visualize a potentially profitable, vacant market niche that corresponds with their technical skills, passion, and business savvy. You, as a potential entrepreneur, might tolerate the risk to go out on your own and start from nothing, since your excitement, expertise, and determination could drive you to succeed. If you have the talent and the temperament and are relatively young when you do this, you may not have a lot to lose.

Since you may want to start a business at some point in your career, your *Ready for Takeoff!* book needs to bring this career option of entrepreneurism to life. Therefore, this chapter will feature input from three very successful entrepreneurs: Wilson Greatbatch, a world-famous inventor; Tim Klein, an electrical engineer who, at age 26, co-founded a high-tech company that is now globally successful; and Susan McCartney, the director of a small business development center who guides budding entrepreneurs. By the end of this chapter, you will know a lot more about being an entrepreneur from three different perspectives.

WILSON GREATBATCH

Wilson Greatbatch is an ultimate inventor, entrepreneur, and American hero. In 1960, he invented the first successful implantable cardiac pacemaker. This single invention is estimated to save 500,000 lives per year and has grown into a multibillion-dollar business. He has founded multiple companies including Wilson Greatbatch Ltd. Throughout his career he has been self-funded. His ever-expanding knowledge, creativity, and determination to help humankind have allowed him to achieve more than 325 U.S. and foreign patents. Dr. Greatbatch holds BS and MS degrees in Electrical Engineering along with seven honorary PhD degrees. He is a member of the National Inventors Hall of Fame and the National Aerospace Hall of Fame. In 1990, Dr. Greatbatch received the National Medal of Technology from President George H. W. Bush. On his 72nd birthday in 1991, he traveled 162 miles in a solar-powered canoe of his own design. You'd think, with all this, Dr. Greatbatch would have retired long ago. No way! Now, in his 90th year, he remains filled with passion to solve our most pressing and technically complex problems. It is an honor to feature Dr. Greatbatch in this chapter and to summarize his career. Notice especially how Dr. Greatbatch's thirst for interdisciplinary technical knowledge has allowed him to grasp technology well beyond his engineering major and solve vitally important problems. To every engineering student in America: Wilson Greatbatch invites you to follow in his footsteps or walk alongside him as a partner. There are problems to solve, work to be done. These challenges mean opportunities for you.

The Implantable Pacemaker

Prior to Dr. Greatbatch's invention, countless people died of heart failure, in particular due to blockage of the heart's electrical conduction system. As a young research and medically minded electrical engineer, Dr. Greatbatch was experimenting with an oscillator to record heart sounds. In the course of his experimentation, he came upon a circuit that could regulate the pace of the heart. Inspired by this discovery, Dr. Greatbatch converted his barn into a laboratory, investing his own $5,000 to refine his invention. Success! In 1960, after two years, he patented his handmade, world's first successful implantable cardiac pacemaker. How important was this breakthrough? In 1983, the National Society of Professional Engineers selected Dr. Greatbatch's implantable cardiac pacemaker as one of the ten greatest engineering contributions to society during the previous 50 years. Eight years later, in 1968, he developed a lithium-iodine battery that could be hermetically sealed to power the pacemaker. This battery could be implanted in the human body without harm and give power to the heart for 10 years—another incredibly important breakthrough.

The general nature of this book does not allow detailing Dr. Greatbatch and all of his inventions. The invention of the pacemaker has just been described as an example of what inspiration followed by hard work can accomplish. Thomas Edison once said, "Genius is 1% inspiration and 99% perspiration." Dr. Greatbatch has continued throughout his career to broaden his interdisciplinary engineering skills, including reaching into biomedical technology. He set up a company and secured four patents on a cure for AIDS. Now he's deeply into solving our energy crisis.

Energy

Big problem, great opportunity for solutions—Dr. Greatbatch believes that we are going to run out of reasonably recoverable oil within the next 40–50 years. He has believed this for the past 10 years, and it has been his passion to come up with solutions for alternative energy sources. Here is a summary of his three incremental steps required for future energy consumption:

1. ***Diesel Vehicles.*** The United States should change to diesel-powered vehicles that are not powered by petroleum—including ethanol, which is an additive to gasoline. Europe is far ahead of the United States on this score, since they do not have our oil resources. We can produce 17 gallons of oil from one acre of corn, 40 gallons of oil from one acre of soybeans, and 100 gallons of oil from one acre of canola grain. This vegetable oil is a renewable resource, and only diesel engines can run on it. Dr. Greatbatch and his team have been working toward an ideal mixture of these vegetable oils that could be an effective, environmentally green fuel for vehicles.
2. ***Nuclear Fission.*** The Chernobyl disaster and Three Mile Island nuclear incidents have frightened many people from this option. France, however, has provided 70% of its energy needs with nuclear fission plants. France operates these plants safely under the supervision of nuclear engineers. The United States will have to do the same, along with solar, wind, and geothermal power. Building nuclear fission plants is a necessary intermediate step to meet our energy needs. Nuclear fission does have a downside of radioactive waste; therefore, it is an intermediate solution.

3. *Nuclear Fusion.* This is the ultimate solution. The sun burns by nuclear fusion, that is, by fusing the very light elements of helium and hydrogen. A workable nuclear fusion system of energy is a long-range project; it will take at least 60–70 years to develop. We need to create a sustainable nuclear fusion process that equals the temperature of the surface of the sun. That won't be easy, but neither was the idea of flying to the moon when President Kennedy announced the goal in 1961 to put an American on the moon within 10 years. We beat that schedule by two years when Neil Armstrong set foot on the moon in 1969. Defined goals are extremely powerful!

How do the sun, the moon, nuclear fusion, and Dr. Greatbatch come together? Here's an idea from Dr. Greatbatch: helium 3 is an ideal ingredient for nuclear fusion. The problem is there is no available helium 3 on Earth. Helium 3 does come toward Earth on the solar winds, but it deflects off the Earth's atmosphere and it lands plentifully on our moon. Dr. Greatbatch concludes that after we establish a moon colony, we can mine helium 3 and liquefy it for a trip back to Earth on a spaceship. He estimates that nuclear fusion from a single 25 ton liquefied payload of helium 3 from the moon will meet the energy needs of the United States for one year! Dr. Greatbatch sees the nuclear fusion solution to be 40–70 years away, but that does not deter him from working on it right now. He has, at his facility, a 10,000 pound stainless steel chamber where he is trying to ionize helium on the way to potentially fuse helium atoms. Dr. Greatbatch has been a staunch supporter of university research in pursuit of inertial electrostatic confinement fusion. All this on the way to nuclear fusion on Earth for ultimate energy! It's going to happen within 40–70 years; just ask Dr. Greatbatch! Then Dr. Greatbatch asks a group of engineering students to consider, "Do you think I am the one who is going to make a nuclear fusion breakthrough in 10, 20, or 40 years? How about you?"

Wilson Greatbatch's Wisdom for Engineers

- Engineering education is extremely valuable with many opportunities within and beyond the engineering profession. For example, medical schools now will accept an engineering degree in place of a premed major.
- Become interdisciplinary. To reach your career potential, continually learn technology beyond your specific major. We have to learn the interdisciplinary language of fields other than our own.
- Don't fear failure; don't crave success. Never avoid doing something because you fear it won't work. You shouldn't look only for success or peer approval.
- Nine out of 10 things I try don't work. The 10th one pays for the other nine.
- If you run into a problem that's difficult and you solve it in an eminently satisfactory way, don't worry about the cost—people will pay the cost.
- Your work should be an act of love. Just do it because it's a good thing to do.
- Prepare to patent new technology that you're working on. Keep a bound ledger, a laboratory notebook in which you describe in detail an invention

that may be patentable. Include a written description and drawings, all in ink—not pencil. Then, sign and date the page along with signatures of two individuals who are not co-inventors. Do not write anything on that page again. Take your bound ledger to a patent attorney who is an engineer to have your patent application processed. Your attorney can do a patent search to ensure your idea is new. Google also has a patent search engine.

- Don't worry too much about secrecy. I like to share my ideas with others. Talk with each other and with experts about ideas that you are working on. I would never have invented the pacemaker if I didn't talk with medical people and learn about heart blockage. Medical doctors explained this to me so that as soon as vacuum tubes were replaced by transistors, it was feasible for me to use electrical engineering principles and invent an implantable cardiac pacemaker.
- Learn about your area of expertise and expand your knowledge by reading technical journals and attending technical conferences. Talk to the experts and share your knowledge. Have a never-ending thirst for new ideas that can help you make a positive impact on the world. That's what a meaningful life is all about!

TIMOTHY KLEIN

Timothy Klein received a BS degree in Electrical Engineering in 1984 and distinguished engineering honors thereafter. He is CEO of ATTO Technology, a leader in high-performance computer storage products that he co-founded at age 26. We are privileged to have him in this chapter to relate his personal story of entrepreneurism. Mr. Klein will describe how ATTO grew from the dream of two young fellows in their mid-20s to being a global high-tech leading corporation 20 years later.

Are You an Entrepreneur?

Mr. Klein opens discussion with a group of engineering students by asking, "How many of you have sold anything on eBay?" Some hands go up. To that he responds, "Congratulations—you are entrepreneurs. That's the same kind of thing an entrepreneur does—marketing and selling your own products."

Now, Mr. Klein will tell you about his company, then his own entrepreneurial journey.

What Is ATTO Technology?

ATTO is a prefix that means 10 to the 18th, that is, one quintillionth or a millionth of a millionth of a millionth. Mr. Klein acknowledges that this name, ATTO, tells you that the founders are engineers. ATTO Technology now is used in 98% of all movies and television shows that are produced in the world. They make the interface cards that plug into computers that attach to disk storage. ATTO is one of the very few companies in the world that provide the technology that allows production companies to stream media to the computers fast enough that people can create very high definition environments.

ATTO products are used for mixing all of the various sounds and visual effects that are used for a movie. It's an exciting, fast-moving business that provides networking data storage equipment for a number of different technologies. Their customers include Pixar, Disney, Sony, Panasonic, KVC, and Dolby, to name a few. ATTO now has 135 employees, who are mostly engineers.

Tim Klein's Entrepreneurial Story

Decision: Be an Engineer. Mr. Klein's steps to being an entrepreneur began in high school, as he evaluated what career field to study when he went to college. One of the things he did was to look up the Fortune 500 companies, where he discovered that 70% of Fortune 500 CEOs had engineering degrees. Once he became an engineering student, he found that he liked his choice of study. Moreover, he realized that engineering students have more options than he had imagined: lots of opportunity within the field of engineering itself, as well as opportunities to be a medical doctor, a pilot, a financial analyst—the list goes on.

Corporate Experience First. During college years, Mr. Klein was an intern at Motorola for two summers and joined them as a full-time employee after graduation. He worked for Motorola for two years doing computer hardware engineering. Then he joined Barrister Information Systems, a company specializing in sophisticated technology for law office support. Mr. Klein was working on hardware development for new CPU processors when he met an engineer who was working on software. Together, they added more and more CPUs into the computer system to the point that everything was backing up in the bin. Problem? Potential opportunity!

Inspiration. They got to thinking: if we can figure out how we can accelerate the disk drive, we can help not only this company, but a lot of companies because there's this big problem in the marketplace that no one is addressing—that is, how do you accelerate disk drives? That's where they came up with their first product idea. Mr. Klein and his co-founder left Barrister in April 1988 and decided to go into business.

The Launch. ATTO's first corporate headquarters was in Mr. Klein's family room. No extra dollars required for that "home office," but what about the expenses involved in starting a business? Neither Mr. Klein nor his partner, each at age 26, had accumulated any significant money. He had a small house without any equity, barely able to meet the mortgage payments, and drove a 12-year old Dodge Dart. The good news, says Mr. Klein, is that they had very little to lose if they failed! But starting a business requires capital—money. What to do?

- ***Banks?*** Two 26-year-old engineers with no money, no business management experience but, they believed, a hot idea. Well it's worth a try—isn't that what banks are for? They approached banks, asking for venture capital loans, and as Mr. Klein says with a joking smile, the bankers "opened the vaults and said come right in—take whatever you want." Mr. Klein shakes his head, still smiling and says, "Yeah, they were just throwing money at us." Not!

- *Credit Cards.* Mr. Klein absolutely discourages other people from doing this, especially today, but he admits that he ran up $65,000 of credit card debt to get the business going, based on the conviction that their idea was going to become profitable.
- *Scrap Parts from Companies.* Mr. Klein and his partner got started on their equipment inventory by going around to manufacturers and begging for any surplus parts that would be useful to infant ATTO. "We accepted circuit boards, de-soldered some of the capacitors, took processors off of things—we literally built our first prototypes out of nothing but scrap materials."
- *Family and Friends.* They were a lot more accommodating than the bankers. Mr. Klein and his partner sold shares at $500 apiece, then after seven years paid them back at $15,500 per share. Let's see: a return on investment (ROI) of 31 times their initial investment in seven years. Not bad. Family and friends, you were richly rewarded for significant risk. Most start-up companies are not as successful as ATTO, but bankers, you missed a winner. The credit card debt was paid in full, and Mr. Klein does not normally believe in accruing credit card debt, especially at current interest rates!
- The successful use of money from family and friend investors, along with money from development agencies who provide seed funding for promising start-up businesses then allowed ATTO to qualify for a loan of additional capital from a bank.
- Eight months after starting the business, they moved corporate headquarters from Mr. Klein's family room into the University at Buffalo's incubator for start-up businesses. Later, business expansion required them to move to several other buildings.

Today, ATTO is located in beautiful new buildings that house R&D, engineering, and manufacturing of the products that Mr. Klein described earlier. ATTO is now doing business in the United States, Canada, Mexico, Central and South America, Australia, New Zealand, China, Korea, India, Pakistan, Taiwan, Africa, the Middle East, Europe, and Japan. On July 15, 2008 ATTO had a big 20th anniversary celebration reception for employees, customers, and the community. A person attending the event would find it hard to believe that this thriving international corporation was only 20 years removed from two young guys with an idea in Mr. Klein's family room.

Tim's Tips for Engineering Students

- *Big Company Experience.* When I graduated, I wanted to work for a big company. There are pros and cons to working for a big company. You get to see how a large company operates, the politics, the teamwork, and how to get work done in a large organization. You work on a project with a team and develop specialized capabilities with a limited view of the entire operation. In a big company you get to do neat things with a lot of resources.

- ❖ ***Small Company Experience.*** A smaller company taught me how to work across a wider spectrum of functions: production, operations, marketing, sales, engineering, and so on. It was a broader view of the world—defining what all the departments are and what they do. Meanwhile, a smaller company does not have the depth of resources that a large one typically has.
- ❖ ***On Your Own.*** When you start your own company, you have basically nothing. Small company experience was very helpful, because when I went into business, I had to have a grasp of all the functions of a successful company and ability to carry them out. You're developing, designing, manufacturing, and selling. You're also the support person, the person handling finances, and the one hiring the right people. It's exciting; you're never bored—there's something new happening every day.
- ❖ ***Two Ways to Grow a Business.*** These include:
  - Bootstrapping involves a very small investment. You take your products out to market, get revenue flowing, and reinvest the profits to grow the business.
  - Using other people's money through venture capital companies that invest in start-up entrepreneurs who appear to have a great idea and a favorable probability of succeeding.
- ❖ ***Conservative Risk for Engineers.*** I've told you a lot about starting a business that involves risk. Meanwhile, I understand that we engineers are conservative. Actually, I consider myself to be extremely conservative. I take educated risks, not long shots. I study what I do, pay attention, and calculate what the odds are for the various options. In general, I use a more conservative decision strategy. That's the engineering approach to being an entrepreneur.
- ❖ ***Business Plan and Customer Focus.*** Have a business plan, update it each year, and focus it on the needs of customers. Research the customer and sell your concept or product based on their need. You can have a great technical idea, but if there's no customer need, the idea won't go anywhere.
- ❖ ***Believe in Yourself, but Listen to the Counsel of Others.*** We engineers may think we have it all figured out, but it is important to get input from other professionals. Attorneys and accountants tend to be conservative. They will tell you what not to do, but listen to them. You need to understand all dimensions of a business issue. Any business executive, including an entrepreneur, needs to balance risk and caution. If you take no risk, you won't be in business, so don't be discouraged by your family and friends who say "you're nuts to do this"—especially if you have a well-founded belief that you will succeed.
- ❖ ***When to Involve Others (Non-Engineers).*** I made a mistake at the beginning by focusing too much on technology and not enough on the business. If I had involved others in sales, for example, the business would have taken off faster. How do you know when to add others? When you can step

back and say, "I'm probably not the best person to do this and can afford to bring someone else in."

❖ ***How to Get a Business Person to Invest in You.*** There are several ways:
  - Some people are very good at selling a concept: on paper, in a PowerPoint presentation, and so on. These people might get an initial commitment without going further.
  - Some customers say, "I want you to have a finished prototype, functioning well before I'll touch it."
  - Other customers say, "Finish the scaled up product and when it's working 100%, I may be interested."

❖ ***Sell the Prototype.*** As you do your market research and business planning, you'll identify potential customers who need your product. Then, take your prototype out and shop it around. You might sell your prototype to someone who has the interest and capability of producing it. In that case, they might produce it for you, and you'll have a customer.

❖ ***Enjoy Being an Entrepreneur.*** I've thoroughly enjoyed being an entrepreneur—coming up with an idea, turning it into a business, and watching it grow. Perhaps you see a market need that excites you. If you have the passion and the right combination of technical and business skills, being an entrepreneur might be great for you too. I wish you happiness and success in whatever you do!

SUSAN MCCARTNEY

Susan McCartney is the Director of the Small Business Development Center, located at the State University College at Buffalo. She became Director of the center in 1990. Her center provides counseling to entrepreneurs and small business owners with planning, marketing, financial analysis, human resources, and other necessities of starting a business. The Center has worked with over 15,000 entrepreneurs and small business owners, helping them to get started and remain successful. Ms. McCartney was awarded the title of New York State Business Advisor of the Year in 1987. Ms. McCartney has a BS degree in Biology, an MBA, and is working toward a PhD in Higher Education Administration. She has also been an entrepreneur as co-owner of American Media, Inc. It's a privilege to feature Susan McCartney in this chapter, following entrepreneurs Wilson Greatbatch and Tim Klein. Ms. McCartney will focus on the fundamentals of successfully starting a business and keeping it running.

Basic Information for Starting Your Own Business

Ms. McCartney offers you this information to help you actualize your dreams as a potential entrepreneur.

❖ Entrepreneurs have an extraordinary ability to dream. The challenge is to manage your dreams into reality.

❖ Strategic planning is the means by which you can actualize your dream and stay in business. As a logical engineer, you are better able to understand this rule than others.

Stay focused on these fundamentals and those that Ms. McCartney will now explain to you. Why? If you decide to start a business, you will be more likely to succeed.

Characteristics of Successful Entrepreneurs

❖ ***The Ability to Envision a Different State or Condition.*** Successful entrepreneurs can see things in a different form. Entrepreneurs, especially engineers, are creative thinkers. They are fixers. They can see what needs to be improved. They observe, analyze, and then focus on a particular solution that meets a customer need. They see what a customer has now and how it can be improved. You, as an engineer, are more likely to have this creative awareness, as well as the skills to build a new product or process that can be sold to a customer. As a potential entrepreneur, do you have natural curiosity and creativity as to how a thing or process could be improved? Do you want to do something about it?

❖ ***High Energy.*** Entrepreneurs don't watch the clock. They are in the "flow state" of focus toward having their dream become reality. They keep going and going, with the ability to sustain this energy for years. Have you ever been motivated like this on a project, especially one that you generated?

❖ ***The Ability to Affect Others' Behaviors.*** Call this leadership; call it persuasiveness. Call it the ability to motivate others—your family, team members, customers, investors, and so on—that you have an idea worthy of their support. You will not be able to get someone to buy into your idea without convincing them that you have a winner for them—that you represent a sound investment. You're asking for their money, time, or loyalty; prepare to let them know that it's worth it. Could you do this if you believe in your idea?

❖ ***Risk Tolerance.*** Entrepreneurs are willing to take risks, but only on themselves and an idea they believe in. They are not gamblers. They take calculated risks after the proper analysis of feasibility and a market for their idea.

❖ ***You'll Love It.*** Once you've been an entrepreneur, once you've owned your own business, you're in love with it and are unlikely to ever go back to a conventional employer.

Strategic Planning

Strategic planning is a necessity. In the United States, it's easy to start a business; the challenge for an entrepreneur is to stay in business. That calls for a sound action plan. Note: your strategic plan needs to be written in detail and not just in your head.

Four Key Elements to a Strategic Plan. Ms. McCartney describes the following:

1. *Setting Objectives.* What are you specifically going to do, and with what timeline? Your plan will include both long- and short-term objectives with specific steps to achieve them.
   - Long-term objectives include where you expect to see your business three to ten years from now. There is a big strategic advantage to planning where your business is going to be 10 years from now, because it will influence decisions today. Specifically, it may affect your choice of location, law firm, and so on.
   - Short-term objectives are more quantifiable. They are more focused on timeline, that is, what you're going to do and when. Specifically, during the next 12 months, how many employees will we have? How many customers? How much money do we need to make during the first year?
2. *Marketing.* Marketing is very important. Every entrepreneur needs to have a customer who is willing to buy the product or process. Without customers, a great idea will go nowhere. Entrepreneurs are constantly in search of new customers and maintaining existing ones. Reaching out and touching the right target market—a person who is willing to pay money for your product is essential. Entrepreneurs are excited and anxious to get their idea out to the world. This is natural, and the entrepreneur's enthusiasm is important.

   But, Ms. McCartney suggests entrepreneurs take a step back and do a market analysis before unleashing their enthusiasm. The more market research you do up-front, targeting your customer niche, the more effective and efficient you'll be in your marketing.
   - How do you do your market research? Go to the library and research models of the business you would like to start. You may be able to get the names and contact information of people who may be willing to help you. It is important to know who else might be in your proposed business.
   - You must do a competitive analysis. You have to know what the competition is doing. Ms. McCartney's Small Business Development Center and others like hers can assist you in gathering competitive information, target market information, demographic profiles, and other important data.
   - Once you and your counselor review the research information, you develop a marketing communication plan. Entrepreneurs love this, because, having done your research homework, you will broadcast your great idea to the right market niche of customers. You'll be thrilled to see your message in a trade journal, newspaper, or whatever medium is appropriate.
3. *Human Resources.* As needed, build yourself a very good team; take the time to make good personnel decisions. Your human talent base is critical to your success. Entrepreneurs are very confident, to the point

that they underestimate the importance of others, especially at the start, and that includes lawyers and accountants to give essential advice. Be sure to have legal employees as opposed to paying them "under the table"; it is a misdemeanor to pay people off the books.

4. ***Financial Analysis.*** Ms. McCartney says that money is the first thing that people starting a business want to talk about. The money element is very important, but business advisors such as Ms. McCartney will address finance last. Well, now's the time; show me the money! How?
   - ***Business Plan.*** Having a sound business plan is the first step to getting the money.
   - ***Understand the Realistic Business Costs.*** You need to know your start-up costs as well as the expenses of running your business for a year. The main reason entrepreneurs failed within the first three years is that they didn't have enough money to sustain them when they didn't have enough clients. On a positive note, you have a good chance of success if your business can stay alive, continuing to market your product to the right customers, for at least a year without pulling back on your plan.
   - ***Your Start-Up Money.*** Figure out your start-up expenses and your operating costs for the first year. Once that is determined, that's the amount of money you're going to need to raise. So, where do you get it?
     - ***Individuals Who Know You.*** This is the number one source of money for a business starting from nothing but your idea (as opposed to buying a business or a franchise). Ms. McCartney suggests that the best way to get money from them is to show them your business plan. Ask them to take a look at your plan with the idea that they might be an investor. You're inviting them to invest, rather than borrowing money from them. Have legal counsel to be sure you're doing an investment contract correctly. Ms. McCartney observes that your mother is the most likely person to loan you money or invest in you. In fact, mother may not tell anyone else in the family she made the loan—it's hush money. Ms. McCartney says that Mother's Day could also be called Small Business Day. Just don't take advantage of your mother unless you have a business plan that will work.
     - ***Development Agencies.*** As a potential entrepreneur, be glad that you are an engineer. A development agency is more likely to favor an engineer than anyone else. You have the technical skills to create a product that will sell. You're also likely to have the analytical mind that can create a logical, disciplined business plan.
   - ***Your Financial and Legal Reputation.*** It is very important that you have a good credit rating. That includes a record of always paying your credit card balance on time and being free of any bankruptcies or felonies in your past. Your credit and legal ratings will be checked before any financial institution loans you money. Generally, you will not get start-up money from a bank.

Summary of Entrepreneurism

We are privileged to have the contributions of these very successful entrepreneurs in this chapter. Wilson Greatbatch, Tim Klein, and Susan McCartney have extended powerful advice to you if you ever choose to go into business for yourself. And who better than an engineer will contribute more to our country and the world by starting a business, building something that will solve a problem, meet a customer's need, and create wealth?

For more information about how to start your own business, Ms. McCartney suggests you visit the Web site of the U.S. Small Business Administration, www.sba.gov. The Web site is titled "Programs and Services To Help You Start, Grow, and Succeed" and has free online training and information about marketing, finance, and every aspect of starting a business. Best of success to you future entrepreneurs!

Chapter Summary: Takeoff Tips

Wilson Greatbatch's wisdom for engineers:

- Engineering education is extremely valuable. There are many opportunities.
- Become interdisciplinary. Learn technology beyond your specific major.
- Don't fear failure; don't crave success.
- Your work should be an act of love; do it because it's the right thing.
- Prepare to patent the technology that you're working on.
- Don't worry too much about secrecy. Share interdisciplinary ideas.

Tim Klein's tips for engineering entrepreneurs:

- Big-company experience is good for developing teamwork/specialized capabilities.
- Small company experience is good for working across a wide spectrum of functions.
- On your own you grasp all functions of a successful company and carry them out.
- Two ways to grow a business: (1) bootstrapping and (2) using other people's money.
- Take conservative risks, by studying and calculating the odds for success.
- Have an up-to-date business plan that is focused on the needs of customers.
- Believe in yourself, but listen to the counsel of others.
- Know when to involve others who are non-engineers.
- Know possible ways to get a business person to invest in you.
- Enjoy being an entrepreneur!

Susan McCartney's characteristics of successful entrepreneurs:

- The ability to envision a different state or condition.
- High energy: don't watch the clock; keep going and going toward success.

- ❖ Risk tolerance: willingness to take risks, but only on themselves and their idea.
- ❖ They love it, and are unlikely to go back to a conventional employer.

Four key elements to a strategic plan:

1. Setting objectives: what are the objectives and with what timeline?
2. Marketing: customers needed. Do market research and analysis up-front.
3. Human resources: talent is critical for success. Make good personnel decisions.
4. Financial analysis: have a sound business plan and understand realistic costs.

Exercises

1. How could you be an entrepreneur?
2. What is Wilson Greatbatch's dream as the ultimate energy solution?
3. Review the section, "Wilson Greatbatch's Wisdom for Engineers." What advice will you use?
4. How did Tim Klein decide to be an entrepreneur and found ATTO?
5. Describe three of "Tim Klein's Tips for Engineering Students" that you will use.
6. Explain Susan McCartney's "Characteristics of Successful Entrepreneurs."
7. Explain Ms. McCartney's four key elements of a strategic plan for entrepreneurs.

CHAPTER 18

Academic Careers and Graduate School

Should I go to graduate school? How about a career as a professor? As an undergraduate engineer, you may be considering graduate school as an attractive option after you receive your bachelor's degree. You may be thinking about a master's degree or PhD in engineering, perhaps an MBA. If you go all the way to receiving your doctorate, an academic career can be very rewarding.

The subject of graduate school and academic careers is important for you to understand as you continue to build your optimal foundation for your career. You need to understand the advantages of graduate school and possibly an academic career after obtaining your bachelor's degree.

HARVEY STENGER

Dr. Harvey G. Stenger Jr., the Dean of Engineering at the University at Buffalo, the State University of New York, will give you insights about different graduate school options, as well as thoughts about an academic career. Dean Stenger received his BS degree in Chemical Engineering at Cornell University and his PhD in Chemical Engineering from the Massachusetts Institute of Technology. Prior to coming to the University at Buffalo, Dean Stenger was Professor and Dean of Engineering at Lehigh University. He has outstanding credentials and perspectives to share as you consider your future in graduate school and the possibility of an academic career.

Dean Stenger can relate to the importance of undergraduate work experience. During summers after high school and his freshman year at Cornell, Dean Stenger worked as a surveyor for an engineering consulting firm. He also worked for two periods as a co-op student. Rounding out his non-academic experience, Dean Stenger has served as a part-time consultant for 20 companies during the last 25 years, keeping him abreast of the private sector business world.

YOUR PLANS

As you proceed through your academic program and work experience, you may have occasions that provide some insight as to where you would be happiest and most productive in the future. Recognize these "ah-ha" moments so you can act on them in a way that will be best for your future career. For example, in Dean Stenger's case, he got a sense during his co-op employment, that he might be more intellectually fulfilled with more control over his life in an academic career. With this in mind, he talked to a professor who encouraged him to pursue his PhD, and then become a tenured faculty member at a university. At that point, the professor advised, "You would have a lot of intellectual stimulation and professional freedom." So, chemical engineering student Harvey Stenger considered his personal thoughts and the professor's advice and decided: "I'll do that. I'll go to graduate school and become a professor." He followed through with his decision: he went to MIT and earned his PhD in chemical engineering. He then became a professor at Lehigh University, where he was able to work on anything he wanted to do. Later, he became Dean of the Lehigh Engineering School with the challenge of balancing administrative responsibilities along with his academic interests. That summarizes Dean Stenger's decision and progress into his academic career. Now let's return to you.

Dean Stenger asks you a good question: What's your plan? A person who hasn't given planning much thought might respond—you mean for tonight or this weekend? Having read the previous chapters of this book, you know that Dean Stenger is asking you for your academic and professional plan, based on your abilities, interests, and life goals. Dean Stenger's thinking about career planning parallels that of our other successful experts from industry: to make the best of yourself and your future, you better have a plan. In fact, force yourself to have a plan.

Lots of Options

Dean Stenger asks a group of undergraduate engineering students: how many of you plan to pursue a PhD? How many of you plan to go into industry after getting your BS degree? How many of you haven't decided yet? Different hands are raised after each question. Then Dean Stenger states a number of other options for after receiving your bachelor's degree. You can pursue an MS in Engineering, an MBA, an MS in Education, an MS in Finance, Medical school for an MD, Law school for a JD, or a licensed Professional Engineer. The list goes on; the point of emphasis is that a BS degree in engineering is a superb foundation for continuing your education, whether in engineering or in numerous other fields. Your study of engineering will, in fact, provide you with many professional options. But now, since this is an engineering career launch book, let's have Dean Stenger tell us about engineering in the future.

THE FUTURE OF ENGINEERING

It is much more complicated than it was 10, 20, or 30 years ago. We have come a long way since slide rules and doing calculations by hand. Now we are developing microelectronic devices that can reside inside people's bodies, wandering around inside your intestinal tract and sending data to a computer that monitors your health. We are creating artificial organs for people. We are modeling eyes to try to prevent blindness. We are trying to understand the entire Earth and its environment in the future. These are much more complicated problems than we were solving 20 years ago. But these are important problems—and opportunities—that we have worked our way up to through continual technical progress.

Commonality

What do all these problems and opportunities have in common? Do we need to learn more physics? More biology? More chemistry? There is a lot of research that lies ahead of us, research that may take years before we can specifically identify the problem. And that research won't be confined to one discipline. As we learned from Wilson Greatbatch in the last chapter, technical breakthroughs are made through interdisciplinary knowledge. Breakthroughs will be made by teams of highly educated engineers bringing different disciplines with different perspectives to a problem.

Exciting examples of research in the future: the potential of the whole world's ecosystems being modeled on a computer. The human brain will be modeled on a computer. Human behavior and social networking will be modeled, with one possibility being the identification of potential terrorist groups. Future research will be very simulation oriented. We will be living in a matrix world—a simulated world of understanding of what's going on and why.

Graduate education is needed to solve these complex problems. You cannot be at a technical level to be at the cutting edge and doing research in these problems with a four-year engineering education. The future of engineering will be more complex than a person holding a bachelor's degree can handle.

A bachelor's degree in engineering is an excellent foundation for a career in management. Dean Stenger predicts that in years to come, a BS in engineering will become the dominant pre-management degree. That is because management will require an understanding of technology. Preparation for management will be strengthened with an MBA in addition to your bachelor's degree in engineering.

Engineers in the Future

Eventually, the engineering profession will require advanced engineering degrees. Why? Requiring advanced degrees is where medicine and law are right now. The same is true of health science such as physical therapy. In each professional field there is more and more to learn in order to practice as a professional. Engineering will be included in those categories, especially in view of its rising importance. So, more advanced education and credentials have become a commonplace requirement. Can you think of a profession other than engineering that you can now practice with a four-year bachelor's degree? How about accounting—but don't you need a CPA license to be regarded as a professional? Dean Stenger predicts that in the future engineers will have a PhD and will be compensated accordingly. You may want to give that some thought, since you'll be around 50 years from now, and especially, if you really want to play at the high end of engineering.

What, in Particular, Do You Want to Study?

This question is important since the answer can be a big step in preparing you for your career. Give thoughtful consideration to the self-evaluation that you did in Chapter 3 regarding your strengths, interests, values, and goals. Then, talk with people who are in your possible fields of interest. If you're headed for a co-op assignment, plan to connect with engineers at various levels and functions. They might give you their perspective on the best academic preparation for whatever career you have in mind. Are they happy with what they're doing? Do they have any advice for you? If you're on a co-op job, remember that you were hired to do work, not get career counseling, but there will be opportunities at lunch or other appropriate occasions where you can discreetly obtain good information. See if you can connect with someone in research and development (R&D) as you contemplate an MS or a PhD in engineering. Meanwhile, if you are on a co-op assignment you will, as did former co-op student Harvey Stenger, be able to assess your future in this type of work based on your ability, interest, and fit.

Discover Your Passion

Throughout this period of your life, in school and work assignments, try to figure out what really turns you on. Maybe it's energy—perhaps making a comeback in nuclear power or another energy source. Maybe you want to solve chronic and aging diseases. Maybe you want to make a smart pill that goes into people's bodies and takes a lot of data that records their health statistics into a computer. Maybe you want to develop artificial human organs. Every engineering discipline calls for creative, highly educated engineers who can advance the state of technology and create businesses. Where are you really interested in make a lasting, positive impact? That and your ability will determine your future course of study in graduate school. In turn, your education and passion to make a lasting professional contribution can be fulfilling and meaningful all your life and beyond.

GRADUATE SCHOOL OPTIONS

Dean Stenger mentioned many options in engineering, business, and other professional fields. Now, he will address some of the options in more detail, especially your action steps toward your goal. This is a tough decision. Dean Stenger reflects on discussions that he has had with many students. Some say,

"I want to go to work for a while and figure out what I want to do, then come back to grad school." Sounds logical, but often it's easier said than done. It might be difficult; you may have a significant salary and car payments, perhaps you have bought a house, might be married, or maybe had kids. You can't predict your future situation several years after your bachelor's degree. That's why Dean Stenger's bottom-line advice is: if you're planning to go to graduate school, think about doing it sooner rather than later. Here are some graduate degrees you might pursue.

Master of Science Degree (MS)

You can get a master of science degree full time on campus or through distance learning from very reputable universities. Distance learning is attractive from the standpoint that you can login, do your coursework from your office or home, never personally see a professor, and get your master's degree. Personally, Dean Stenger advises against this approach unless necessary. Why? Distance learning entails less separation from your normal work environment and less opportunity to immerse yourself in your advanced study. You will be glad that you achieved an advanced degree and the knowledge that goes with it, but Dean Stenger suggests going full time if you can for more total focus in your field of study. As a full-time graduate student, there is a decent chance to get a teaching or research assistantship, perhaps a part-time job on campus, to help cover the year and a half costs of getting a master's degree.

Master of Business Administration Degree (MBA)

Dean Stenger has some successful friends who, after receiving their BS degrees in engineering, went to work for a few years in a corporation or on Wall Street, then returned full time to get their MBA at a prestigious business school. From there, they cut the ties with their previous employers and went on to a variety of business positions. This could mean entering corporate management, becoming an entrepreneur, being a financial analyst on Wall Street, or other options. You could also receive your MBA education at night, with your company paying all or part of the costs. However, Dean Stenger advises going full time, if possible, for the same reasons stated for the MS in engineering.

Doctorate in Engineering Degree (PhD)

If you're going to pursue your PhD, it's good to make that decision by your senior year. As Dean Stenger indicated earlier, a co-op or internship assignment after your junior year may help you make up your mind about pursuing the highest level of engineering education. Talk with some of the engineers at various ages, in different functions, including R&D engineers with PhDs. Are they happy? Do they like what they're doing? Can you see yourself doing what they are? Would you like to get more education to function at an elite level of engineering?

When to Decide

In the fall of your senior year, as soon as you return to campus from your co-op assignment, Dean Stenger suggests that you develop your decision about higher education by taking campus interviews from companies and, if possible, going on some plant visits. That may help you decide whether you want to go right into industry or pursue higher education. Be aware that companies start interviewing on-campus in September, so it's good to get an early start with your resume and interviews, including attendance at technical career fairs completed by mid-November. Ideally, you will narrow down your options and take some plant interview trips during December–January winter break.

GRADUATE SCHOOL: THE PROCESS OF APPLYING

While interviewing for employment, you will be applying to graduate school. The process of applying to graduate school is multi-step.

- *Choose Your School.* Choose the number of schools you will apply to. That should be about four good prospects for you—not ten. How do you narrow it down?
- *Research Your Interest.* This is going to be primary. Have you thought about your passion—what you want to work on? You want a university and department that suit your abilities, interests, and professional goals.
- *Quality of School.* How highly is the school ranked? You can look at the U.S. News survey or go online to see the ranking of schools. That information will be helpful because it will also tell you the average Graduate Record Exam (GRE) score of those who are admitted. That gives you a benchmark as to where you fit when you take the GRE. You can Google "graduate school rankings" as well as "graduate record exam" for details on each subject.
- *Geography.* Consider a place where you would like to live for 2–5 years.
- *What Schools Should I Apply To?* Talk to faculty—your most trusted faculty member(s), including your faculty advisor. You might say, "I'm thinking of going to graduate school and I might work on _(x)_ problem; what are the best universities for this? Who are the best professors out there for that technology?"
- *Letters of Recommendation.* You should understand that in making the above inquiry about possible schools you'll not only get valuable information, you'll also be preparing three professors to write letters of recommendation for you. Your selected professors should be from your department and who know you fairly well. They should feel respected that you're coming to them for advice, and hopefully you can get them positively involved in helping you get into graduate school. Specifically, after getting their advice you'll be coming back to them in November and asking them to fill out a recommendation form for your targeted schools that meet your objectives. You'll need three letters of reference,

so if you don't know three professors well enough, make it a priority to introduce yourself and get to know them. In summary, tell them about your objectives and involve them in your decision, which will help them when they write their letters of recommendation.

- ***Take the Graduate Record Exam (GRE).*** As indicated, you can Google "graduate record exam" for details. There is not a set date for the GRE; Dean Stenger suggests you take it at the end of the summer after your junior year while coursework is fresh in your mind and perhaps after having taken a GRE preparation course. Like the SAT, there are math and verbal sections with each score worth 800, totaling 1,600 points. You will need to score about 730, especially in math—the quantitative part—to go to the best graduate schools.
- ***Statement of Purpose.*** When you fill out your application, know that a very important entry is your "Statement of Purpose." This means that you're going to try to convince the school that you're applying to that you're a good, targeted fit for their graduate program. It's important that you do your homework on this and, ideally, that you know a particular professor at that university who you want to work for whose expertise is relevant to your qualifications and objectives. You'll have a clear definition of what you want to study and why you would be an outstanding addition to the research being conducted at that university. As indicated, all the better if you have a specific professor in mind whose research you can support.

You're ready to take action on the steps outlined in this section to apply to the right graduate schools for your interests and abilities.

AIM AT THE RIGHT TARGET

Now you need to ask yourself some important questions: how good am I? What school will admit me? To emphasize this point, Dean Stenger tells of an exceptionally talented student who got turned down by eight out of nine schools. That student did get admitted to an excellent university and received his PhD, but he applied to eight other elite schools that did not fit his specific qualifications and interests. Therefore, you may want to re-read the Statement of Purpose section above, to be sure that you're aiming at the right target and not wasting your time applying to schools where you don't fit. Talk to your faculty advisor; know where your strengths are. Get your professor's recommendation of the right schools for you. Compare your GRE score with those that your targeted university will accept. There is nothing wrong with calling a possible school and saying, "Here's my record. Am I wasting my time, or do I have a decent chance for admission?" Dean Stenger suggests that it is especially good to visit your targeted school and meet the professor who you would like to work for. Have this professor get to know you, including your qualifications and passion for his or her research. Ask your faculty advisor to call the targeted professor and tell about you and why you would be a good fit.

If you really know where you want to go, that extra research and selling effort is worth it. The bottom line on aiming at the right target is that it's not just about your GPA; your grad school wants someone who will fit in their laboratory too.

There are 180 engineering schools that offer graduate programs. If you're really dedicated to continue your education and you want to go full time, you have a wide range of options. Certain schools will admit you for grad school with a 2.5 GPA. You may not get financial assistance right away, and they may ask you to prove to them that you can make it. To know where your target school zone is, graphing your GPA and GRE scores alongside published university information would tell you where you are more likely to be accepted.

Graduate school is an adventure—have fun. Dean Stenger says that, from his perspective, going for your doctorate can be four of the best years of your life. This is especially true if you pick the right school and a great thesis advisor. Ideally, you have an adviser who trusts you and gives you the freedom to be creative in achieving results. To be sure you get such an adviser, you might talk to some of the other grad students who will tell you which professors are the best to work with. You should be able to make your PhD study a wonderful, enriching experience.

FINANCES

Dean Stenger offers the following information about finances, which is an important consideration. If you're accepted into a PhD program and are a U.S. citizen you'll likely get a tuition and stipend award. In 2009, a typical award would be \$20,000 for 12 months plus they pay your tuition. You ask yourself, "Financially is it worth it? For four years I'm only going to make \$20,000 × 4 = \$80,000." You might speculate that you're missing out on \$40,000 per year since you could get \$60,000 per year if you went right out and got a job. So, given those numbers you would miss out on \$160,000 over four years. You ask, "How long will it take you to make that up after I get my PhD? What's the difference in salary?"

Payback—Return on Investment (ROI)

You can go to a library and look up an assistant professor's salary having just completed their PhD. But, here's the answer: in 2009, the new PhD assistant professor is making approximately \$80,000 for nine months of work. In addition, they are allowed to bring in two months of summer salary, consulting one day per week. Therefore, a 27-year-old PhD assistant professor can make, on the average, \$100,000 per year. Now, if you had been working for four years after your BS degree starting at \$60,000 with 4% annual salary increases you might be up to around \$70,200 in four years. So, after you get your PhD and work as an assistant professor you'll make up the financial difference in a short period of time—in about two or three years you'll pay yourself back.

Summary Perspective

Dean Stenger asks, "Financially, is it worth it? No, that's not the reason you go on for a PhD." The reason you go on for a master's or PhD is that you want to learn more and become smarter. You want to become what Dean Stenger believes the engineer of the future will be, one with advanced skills beyond the bachelor's degree and working at the cutting edge of technology. So, absorb this chapter. Make good decisions. Work hard. Learn your passion and create your future. Have fun. The reward is there for you.

Chapter Summary: Takeoff Tips

- ❖ Make an academic and professional plan, based on your abilities, interests, and goals.
- ❖ There are numerous options after your BS degree, including employment, or many graduate school opportunities in engineering, business, law, medicine, and so on.
- ❖ The future of engineering:
  - Evolving technology presents complicated problems to solve.
  - Teams of highly educated, interdisciplinary engineers will be needed.
  - Exciting research opportunities, especially simulation oriented.
  - Graduate education is needed to solve these complex problems.
  - A BS in engineering will become the dominant pre-management degree.
  - The engineering profession will require advanced engineering degrees.
- ❖ What, in particular, do you want to study?
  - Give thoughtful consideration of your strengths, interests, values, and goals.
  - Talk with people who are in your possible field of interest.
  - A co-op assignment would be a good opportunity for goal assessment.
  - Discover your passion. What technical area turns you on?
- ❖ Graduate school options
  - MS degree: full time recommended or through distance learning
  - MBA degree: full time recommended, perhaps after some experience
  - PhD degree: best to make that application decision by senior year
- ❖ Graduate school: the process of applying
  - Talk to your faculty. Who are the best universities and professors for me?
  - Choose universities and departments that suit your professional goals.
  - Select about four good prospects for your application—not 10.
  - Prepare three professors to write letters of recommendation for you.
  - Take the Graduate Record Exam (GRE) after your junior year.
  - If possible, get to know a desired professor at your targeted university.
  - Your "statement of purpose" on your application should show that you're a good fit for that university and department.

- ❖ Reasons for completing a master's or PhD degree:
  - Learn more and become smarter
  - Work at the cutting edge of technology
  - As a professor, enjoy intellectual stimulation and professional freedom

Exercises

1. Answer Dean Stenger's questions: What's your plan? Why?
2. Why is graduate education needed for future engineering problems?
3. What, in particular, do you want to study? How do you answer this?
4. How far do you want to go? What degree?
5. Describe your key steps in applying to graduate school.
6. How do you aim for the right graduate school for admission?
7. What are the advantages of an academic career?

CHAPTER 19

Becoming a Global Citizen

Contributed by Lester A. Gerhardt, MS, PhD

LESTER A. GERHARDT

Dr. Gerhardt's career combines both industrial and academic experience. His specialty is digital signal processing, and he conducts sponsored research and teaching in this field, emphasizing image processing, speech processing, and brain computer interfacing. He has also done extensive research in adaptive systems and pattern recognition, as well as computer-integrated manufacturing. Following a nationwide search, Dr. Gerhardt was selected as the first chairman of the then newly merged Electrical, Computer, and Systems Engineering (ECSE) department at Rensselaer Polytechnic Institute, becoming the youngest department chair. During his more than decade long tenure as ECSE department chair, the department was cited as the most improved department in the United States by the National Academy of Engineering. In addition, he is a Professor of Computer Science and a Professor of Information Technology.

He has been actively involved in academic administration concurrent with his professorial career. He has held numerous prestigious positions including founding Director of the Center for Manufacturing Productivity, Director of the Computer Integrated Manufacturing Program—the Center was awarded the National LEAD (Leadership, Excellence, and Development) Award, Director of the Center for Industrial Innovation, Associate Dean of Engineering for Research and Strategy, Vice President of Research Administration and Finance, Dean of Engineering, and most recently as Vice Provost and Dean of Graduate Education by special appointment of Rensselaer President Shirley Ann Jackson.

Internationally speaking, Dr. Gerhardt has served as a delegate to the Scientific Affairs Division of NATO and a chairman to its Collaborative Research Grants Program; is a consultant to numerous governments; and is the co-founder of the Global Engineering Education Exchange Program

(Global EEE) in 1995, serving as its founding chair of the Executive Board for over a decade. He is recognized as the originator and architect of the Rensselaer Engineering Education Across Cultural Horizons (REACH) Program—the goal of which is to require an international experience of each undergraduate engineering student, has served on the International Advisory Board of the ASEE, and was recently named as one of only nine faculty in the U.S. designated "agents of change" for globalizing engineering education as part of an ongoing NSF study. He is one of the 19 original framers of the Newport Declaration published in 2009 calling for the taking of deliberate and immediate steps to integrate global education into the engineering curriculum in the United States.

Before going to RPI, Dr. Gerhardt spent 10 years with Bell Aerospace Corporation, where he worked on the visual simulation of space flight including the Apollo Program's moon landing, using both model-based and electronic based imaging technology, and for that he holds several patents and received the Bell Outstanding Management Award.

Among his honors and awards, Dr. Gerhardt was the Inaugural Recipient of the National ASEE Research Administration Award, holds an honorary doctorate awarded by the Technical University of Denmark, and in 2009 was one of only four that received the Distinguished Alumni Award by the State University of New York at Buffalo. Also in 2009, he was appointed as senior advisor to the president of the Institute of International Education (IIE), concurrent with his position at Rensselaer. His focus has been on promoting STEM fields and especially on providing his support for developing countries in Africa, the Middle East, and Central Europe. He is a Fellow of the IEEE and of ASEE and holds several patents.

His diverse professional service includes having served on the NSF Advisory Board, chair of the Engineering Research Council, board member of ASEE, founder and chair of the International Conferences on Computer Integrated Manufacturing, chair of the International Conference on Robotics and Applications, among others. He is currently on the Board of Directors of Capintec, Inc. He remains an active consultant to industry, government, and academe.

INTRODUCTION

Your life already has been influenced and guided by individual choices you have elected to make, coupled with your reactions to unanticipated circumstances and opportunities you have encountered along the way. These include setting educational, career, and life goals and planning for their realization by learning about yourself and about others through proper education and experience. On the other hand, there are uncontrolled circumstances or windows of opportunity that may spontaneously be presented. They may be either passed over or seized with a special passion. Perhaps it's deep insight or just plain dumb luck that a person effectively and successfully joins elements of these two groups together from time to time in one's life. It's in this context that it's important, if not mandatory, for you to intellectually become a global citizen. The opportunity to do so is best embraced as part of the educational process as you prepare for your career. Applicable to all students in their pursuit of any career, this is especially true and necessary in the engineering profession.

Done right, cultivating a proper global attitude and experience will cause your personal life and professional career to become inextricably intertwined to benefit both, teaching you much about yourself and your engineering-self as a result. Hopefully this chapter will provide some helpful hints, but also that it gives you pause for self-reflection, motivation, and guidance.

WHY?

Why is it necessary to have a global attitude brought about by a global experience? The United Nations counts 192 member countries of the world. Adding Kosovo, Taiwan, and the Vatican City brings the number to 195. Besides these, there are dozens of territories and colonies such as Greenland, Puerto Rico, and Bermuda.

Among this large number, one country formed less than two and one half centuries ago attracted a highly diverse population from all over the world, and by the 21st century it unbelievably had grown to global pre-eminence in education, entertainment, sports, science and technology, finance, and government among other fields with only about 4% of the world's population. However, despite its diverse multicultural heritage and composition, despite the fact that 96% of the world's population lies outside its borders, only 20% of its citizens have passports and less than 1.5% of its domestic students include an international experience as part of their education. That country is the United States of America. This individual isolationism is in conflict with the collective "global village" reality of today originally presented by First Lady Hillary Rodham Clinton, and must be changed.

Technologically speaking, our physical world is daily being made virtually smaller by the ever-increasing rapid change of that technology, the very field that you have decided upon as a career. Today we have a technologically borderless planet. The changes of the agricultural and industrial revolutions in the United States pale in comparison to the changes of the information revolution. Hardly more than 100 years ago, we wrote with a quill pen and delivered mail at about 5 mph by horseback using the Pony Express, but today we have e-mail and instant messaging. We first flew for a few seconds at Kitty Hawk in 1903, but only 66 years later men landed on the moon, achieving an escape velocity of 25,000 mph. The computer was invented in 1947, the transistor in the early 1960s, and IBM produced the first PC in the early 1980s. The Web become privatized in the mid-1990s, and here we are today with YouTube, Twitter, Facebook, Google, Amazon, and so on. You just as well could be reading this electronically as in hard copy. Edison said "let there be light" and look what happened, now we have LEDs. Alexander Graham Bell once told Watson to "come here," but

Watson arrives today with cell phone in hand and iPod in his ear. The Boeing 777 was the first paperless design done around the world around the clock. Well, you get the message. Technology is increasing in its rate of change—technology in which you will be a leader will continue to have an even greater impact on society and the environment in which we all globally live. That responsibility rests on your shoulders as well. Simply put, global problems require global solutions. In the United States, the Science and Engineering (S&E) workforce is said to account for over 50% of the GNP of our country, but only comprises 5% of the overall workforce. This is in need of change.

Your environment has become everyone's environment. What we each do in our hometown affects our state, our country, and other countries around the globe. Consider the discharge from the smokestacks of Chicago and the resulting acid rain issues created in the lakes of the Adirondacks in New York. Look at the orange glow around Los Angeles the next time you land at the airport there. Consider the damage pollution has done to the structures and statuary on the Acropolis in Athens, Greece. Remember Chernobyl where our quest for energy and the failure of a nuclear power plant led to the devastation of an entire community, or the Valdez oil spill off the Alaskan coast and the resulting damage it caused. The global list goes on. On a global scale, concerns have been raised about global warming and the melting of the polar ice cap, and holes in the ozone layer, in one way or another reflective of our technological advancement resulting in convenience but also in automobile emissions detrimental to our environmental well-being.

Your energy resources and your usage of them have immediate global impact. The United States with its 4% of the world's population is said to use 26% of the global energy consumption. As countries progress in their development and our demands as individuals increase, the global interdependency increases. In simple terms, we are using current sources of energy faster than it takes to develop, and we must find alternative renewable sources of energy in the near future. This offers another challenge to your generation that must be met on a global basis.

Regarding the societal infrastructures of the countries of the world, it is clear we have evolved quickly to multinational corporations with numerous employees globally located. Offshore production is the norm. The constant search for developing countries for short-term cheaper labor market continues. As the country grows, the labor costs naturally increase and the search goes on for the next in line. All countries that produce any type of products invariably do that for a world market; global import and export are the norm. Truly, we have a global village in terms of economics. The financial crisis of the early twenty-first century showed that what first appeared as a local mortgage crisis was quickly exposed as really a deficiency of all global markets and economies. We were living on virtual monies in a virtual world and playing the video game of life. What seemed to be a Wall Street problem in the United States, quickly correlated to the bankruptcy of Iceland that further exemplified the global interdependencies of virtually all aspects of all of our countries. Such interdependency of all our economies is yet another dimension of our global village existence. Perhaps that will become the silver lining of this century's crisis. It has made both the existence and importance of a global perspective and understanding obvious to all of humanity, and perhaps will provide a bit of recalibration of values for our life and the lives of others on whom we depend, and on our shared residence on this planet on which we live. It is also your moral obligation and responsibility to care for those less fortunate in your own neighborhood, and realize that the neighborhood is now the immediate world.

HOW?

You need to plan to create and develop opportunities to cultivate your passion for global understanding through a planned experience. Succinctly, do as much as you can, as early as you can. This should be a guiding principle in establishing your global perspective. Take all the opportunities that may arise through family and friends to travel as much as possible. Get a personal picture of these wonderful United States through vacations and perhaps family business trips. It's also never too early to experience international travel.

Learn a second language. If a second language is spoken at home, learn it and use it with your parents, grandparents, and so on. Remember most people use it or lose it. Language is most easily learned at a young age when it is learned in the language center of the brain. In later years, language is learned in other portions of the brain which makes it more difficult for adults than small children. A first language is usually learned while we were infants; we first learned to speak, then read, and then write. It is best to learn a second language that same way as when we were young.

Read all you can and use the media to learn about other cultures and countries of the world. Certainly, the Web offers everyone a window on the world, past and current. Use it intelligently to maximum advantage, gaining global knowledge and instilling a passion in you for knowing more and seeing more firsthand.

All that notwithstanding, the greatest opportunities will in all likelihood arise through your educational process; use that channel of opportunity well. For example, even in selected kindergarten classes, children may take special classes to learn a language such as Spanish. In later years, class projects on other cultures are assigned. In high school, families host international students.

STUDY ABROAD

The greatest opportunity for you to gain a global perspective is to go abroad for a reasonable period and learn about other people and their way of life firsthand; usually the first time most of us will have that opportunity is in college.

Inquire as to the availability of international programs. Do they have them? What are they specifically? How do their international exchange programs work? Does participation in the program in any way delay the time to get your degree? Are course credits taken abroad or other international experiences credited to your degree at your home university? Do grades transfer? Is there an additional cost involved? Who are their international academic partners and how does their quality rank? These and other questions need to be asked, and the answers factored into your decision in studying abroad. Here are some personal judgments based on the experience of myself and colleagues and friends that may help you along this part of your life's path.

There are numerous international programs offered by international colleges and universities available via the Web. Usually they have to be arranged between you as a student at your home university and the potential host university directly. This may entail you taking a leave of absence from your home university for a semester, and also individually arranging for course or internship credit approval at your home university prior to going abroad, among other factors. How the tuition at the host university is handled varies and you may find the cost of some such programs additional to your home university's tuition. Basically, such programs require arrangements made between the host university and yourself, gaining supportive approval from your home university academic advisor along with some administrative approvals. As such, international programs that are run and supervised by your home university are generally easier to negotiate and are recommended.

International Experience for Undergraduates

Let's focus on the undergraduate degree for an international experience. Remember: do as much as you can, as soon as you can. International graduate experiences usually can be arranged with your doctoral advisor as you pursue your research on a case-by-case basis.

Timing

In terms of length of time, two weeks anywhere is more of a vacation than a learning experience. There are some excellent summer programs available, and you should explore these when considering electing the option for an international experience as part of your undergraduate education. An ideal time frame for a more immersive international experience is one semester. One academic year is also possible. Most American students prefer a one-semester experience in the spring. This enables better synchronization with the different academic calendars in Europe, Asia, and elsewhere, where most start and end later than in the United States. The late ending is not a major factor since the delay in returning during the summer is not an issue with most students. Moreover, the possibility of an internship abroad following the spring semester is appealing to many. One academic year is also possible and avoids the synchronization problem altogether.

Learn by Doing—International Internships

An excellent way to gain a global perspective is by temporarily working in another country, while pursuing your studies. Such an assignment is certain to immerse you in both the culture of the country and the global nature of your field. Although most such assignments are in multinational companies or local companies where English is the dominant language, you'll be guaranteed to learn the basics of the local language when you shop, order at a restaurant, and just generally live in a foreign country. Such internships also prove valuable as a real demonstration of your openness to new ideas, flexibility, and global curiosity; they are greatly valued by future industrial employers. Make sure to put it on your resume! It is a real education in a multifaceted way. You have a chance to work in your selected field under senior supervision and guidance, and thus learn about the work and how it is conducted firsthand in another country.

I suggest you plan to do a global internship for an extended period so that you have a chance to sufficiently immerse yourself in the culture and the business and to provide enough time to allow you to make a real contribution to the company and realize the fruits of your labor. A semester is an ideal period since it fits nicely into your academic calendar. The summer also permits about three months for such an experience, and is thus ideal for this purpose as well. Most such internships offer a paid position, although some may not, so check it out. Many times it is advantageous to couple an internship with a preceding academic semester abroad. This way, you have infrastructure support by your international host university to help you secure an internship position where it has local ties. Moreover, in some countries, as you apply for such a position from the vantage point of a university in the same country, you encounter much less red tape than you would if you applied directly as an individual from another country. Credit toward your degree must be done in advance with your home university, and most universities will value quality relevant internships by allowing credit to be earned as part of a degree program.

Googling "global internships" is sure to get you an abundant number of sites to explore. Generally, such opportunities to work abroad fall under the following categories: program centric, university centric, discipline centric, and country centric. Some are combinations or hybrids of these basic groupings.

A fine example of the program-centric category is the Global Engineering Education Exchange program. This tuition swap-based program offers either a one or two semester exchange program focused on engineering in almost 20 countries with over 50 international university partners. Each of these options may be coupled with an internship that is handled by the host university with the advantages mentioned before. However, it does not offer a stand-alone internship.

Many universities offer stand-alone internship programs in many countries. Some universities offer it themselves, while others work through external organizations for this function as well as for other international experiences. Consider programs such as Internships–Education, Experience, and Employment (IE3) run by Oregon State University or Global Internships LLC in Dover, New Jersey. Individuals may choose to work with organizations directly, such as Global Services Corporation and the Intrax Global Leaders Internship program. However, please be sure you gain prior approval from your home university before participating. A fine example of a discipline-specific group exchange/internship initiative is the Study in Rome Architecture program done annually by Rensselaer Polytechnic Institute. Many undergraduate colleges and universities offer undergraduates internships in anticipation of their professional studies in law, medicine, and so on, to acquaint the student with the field and practice of these professions. Finally, there are many country-specific international student programs that run the gamut from exchanges to scholarships for study to internships of various lengths. The DAAD (Deutscher Akademischer Austausch Dienst), a well-reputed German organization with offices in New York and other cities, is such a country-based program.

This brief summary should give you some insight into this important dimension of gaining a global perspective and some avenues to explore the opportunities available. Read carefully and heed the advice given, then select one with the approval of your college. Travel safe, learn, and enjoy.

When?

In terms of when to elect to go abroad, the freshman year is too early and generally inappropriate as you have just left "the nest" of your parent's home, and are first beginning to study at a higher level in a new environment. The senior level is not desirable as many spend the last year in specialized courses, seeking opportunities for subsequent employment or graduate school opportunities. That leaves the sophomore or junior year. One might instinctively favor the sophomore year, feeling it is more flexible and feeling the junior year is more specialized in curricula offerings. Others favor the junior year as the student is more mature and self-sufficient. If the level of excellence of the international academic partner is sufficiently high, there should not usually be any problem in getting the equivalent course you desire at the host university for either one or two semesters in the junior year, and therefore that is recommended.

School Quality

Quality should be your main interest when selecting a host school from the choices offered to you. In today's world, the U.S. educational system as a whole remains the best overall. The United States is the largest host country for international students as evidence of that. That said, there are outstanding colleges and universities worldwide that rank equal to or better than most U.S. colleges and universities when considered individually. Look up your home university as well as potential host universities in such annual compilations as *The Top Universities Guide* that includes the world rankings and even the top 100 in engineering for more information. In a recent issue of the top ranked 100 engineering programs worldwide, 19 of the top 50 were in the United States—with the top four being in the United States, and 13 of the second 50 were in the United States. However, that also means about two-thirds of the top 100 ranked universities are outside the United States. Faculty at many of our finest universities spend their sabbatical year at these international universities. Consequently, it is reasonable to assume a junior-level student can and will find satisfactory equivalences that will meet the courses they would normally take at their home university fairly easily. There is an abundance of international universities of high quality that offer a wide diversity of curricula in many disciplines as specialized as normally needed by U.S. students.

Cost

Cost is always raised as a factor initially when an international educational experience is suggested. In a tuition swap type program, students continue to pay tuition at their home university when abroad and pay room and board at the host university. The latter is usually quite similar to the room and board they have been paying at the home university. As a result, the total overall cost is about the same. Of course, the travel to and from the host location is usually the student's obligation, but student fares exist to make this expense more doable. In addition, any travel in the host or surrounding countries is left to the student.

The choice of the international host school to attend from the options offered to the student is usually dictated by the following set of factors: the quality of the school, the offering of the curriculum discipline needed, the language of instruction, and the desirability of the location including concerns for safety.

Language

Regarding the language of instruction, most countries have long preferred to teach undergraduates in their native language, while many graduate-level courses have been taught in English. This is quite understandable and is done as an attempt to retain the culture and heritage of the country through language. However, as the importance of globalization has grown in every country's culture, and as all countries now seek a much more highly diversified student body, there are now numerous courses taught worldwide in English. For example, countries that have fine universities teaching undergraduate courses in English include, but are not limited to, Singapore, Denmark, Spain, France, Germany, Hungary, Finland, France, Japan, South Korea, and India. Total immersion includes an international experience where the student is taught in the native language of the country. This would be truly ideal were it not for the fact that most domestic students in the United States do not have sufficient language capability to be taught in the native language of most other countries. Hence, with the trend to more international instruction in English, this simply requires some refinement in your host school choice, but permits more opportunity to many more U.S. students to engage in an international experience.

Location

In making a choice, consideration may be given to places where a friend has gone before, where a relative might live, or for the quality of skiing nearby, among many other personal factors. In some programs, the student may choose from a set offered or in some the student is assigned. However, the dominant factor in today's world for the student's family, albeit less so for the student, is their concern for safety. This has raised concerns mostly since the events of 9/11. Interestingly, several programs saw an increase of interest to go abroad rather than shrink after 9/11. The Middle East has thus far not been an attraction to U.S. students studying abroad, but nor has Mexico in recent years because of concerns for safety. Hopefully the world will change and both perception and facts dealing with safety concerns will change as well. Until then, be assured personnel at your home university will choose locations that are appropriate in all regards and ones that are safe. Let us all hope that someday this concern can be a non-issue worldwide. It is to that end we are all working and striving for peace everywhere. It is essentially the reason why all U.S. students should have an international experience as part of their undergraduate education.

MAJOR INFLUENCING FACTORS

1. ***Motivation and Passion.*** Be it engaging in an international experience on your way to gaining a global perspective and attitude or some other pursuit of a life goal, it is most important to be self-motivated. More to the point, you need to be passionate about your objective and dedicate your full energies toward that end. Entrepreneurs, Nobel Prize winners, and Olympic champions have all done it—complete dedication to their objective. The most important consideration is that you emotionally and intellectually feel you must be a global citizen and, as such, seek a university that offers you that opportunity. Don't consider it as an educational luxury, but as a necessity for success.
2. ***Learning.*** Learn about your field of interest by diversified formal education as well as experience, and pursue continuous lifelong learning. Technology changes too rapidly to rely on knowledge gained in your undergraduate years only. Learn about yourself and learn about others. Establish friendships and partnerships that raise your level. Think globally and act globally.
3. ***Points to Ponder.*** Be both an engineer and a humanist when you ponder the three questions: Can it be done? How to do it? What is the impact on society if it is done? Most engineers usually focus on the second question—how to do it. For the future, the last question—what is the impact on society if it is done—will take priority and as an engineer you must not only succeed in your objective of reaching the goal set, but take the responsibility for the ultimate social impact of your actions and accomplishments.

4. ***Driving Forces.*** Four of life's driving forces are need, change, imagination, and passion. Use each to your advantage.

   Need is a great motivator. However, each generation seeks to reduce the need of the next generation, offering their children a better life than theirs. In so doing, they reduce the need and thereby the motivational aspect it brings. A critical balance must be maintained of both effects.

   Change is sometimes so slow it is unnoticed or too fast to do much about it. Either way, the reaction comes too late to be effective. The best way to deal with change is to lead it, as the fashion industry does. Clothes don't usually wear out, but become outdated because of the most recent changes in style. Most of all, remember that technological time constants are becoming shorter and shorter, while generational time constants are getting longer with increasing life spans due to the very achievements of our scientific and technological progress. This incompatibility creates challenges as you'll try to implement the latest and greatest engineering discoveries and applications. Success requires a combination of theory and practice. Both are your responsibility. In a global context, the cultural differences will compound these differences and represent another reason for gaining a global perspective.

   Imagination goes hand in hand with vision. Could you have imagined what the twenty-first century would be like 100 years ago and predict our accomplishments? Given your knowledge of that now, what is your prediction of the next 20 years, no less the next 100 years?

   Passion has significant importance—without it you can and will do good work, but you won't be completely content with your life's focus and accomplishments. With it, you can and will do great work and gain leadership roles beyond your own imagination. A person who is passionate about their field is passionate about their life. And remember, you do what you do for yourself as well as for others. Everyone becomes the beneficiary of their own passionate pursuits.
5. ***Leadership.*** A famous CEO of a U.S. company was quoted as saying, "Lead, follow, or get out of the way." But understanding what leadership is and how to achieve it will be a key element of your success, and is addressed in a separate chapter of this book. As it's a major influencing factor in your life, especially when considering the global dimension, it deserves a few words here also.

Think of leadership and teamwork as synonymous. Leadership requires the talent to engage others as active participants and provide motivation, guidance, and methodology for them so as to realize goals and objectives beyond what you acting alone can accomplish. Some important characteristics of good leadership that are often not cited and perhaps undervalued because they are not easily quantified are outstanding people skills in reading people and properly utilizing their strengths to major advantage beyond their individual ability. This is especially important when bridging different cultures with different value systems as in global activities; good basic instincts or gut feelings; both breadth and depth of knowledge without compromising either one; having a fundamental trust, respect, and confidence (TRC) in others; not being overwhelmed by data, but using it only for the information contained within that leads to knowledge and then to wisdom; a holistic approach coupled with attention to detail; and a balance of life and career. True leaders are rarely under stress; rather, they thrive on confronting challenges. Don't confuse a visionary goal-oriented person for a workaholic; they are not the same. An outstanding leader knows the business, the vision and the goals, the metrics, the rules, the people, the strategy for success, and the value of having all this knowledge at their fingertips.

Potential pitfalls include such common mistakes as: letting the theory overwhelm the importance of the practice, and vice versa; mistaking a person's credentials for capability; delegating work without providing proper context and direction for it; not recognizing people who talk the talk but don't walk the walk; confusing a fear of failure with actual failure; and placing excessive emphasis on the process rather than the product.

On a personal basis and most important, never underestimate your abilities. Always try your best to fully realize your potential. Only in that way will you learn your own limits. Be assured that any unrealistic overestimation of your abilities will be quickly adjusted by others.

CONCLUDING THOUGHTS

In conclusion, you'll best pursue life with a passion in all respects. You'll then find that your personal life and professional life will become naturally intertwined, and each has benefitted from the other with minimum effort. You'll note that you'll never say you're going to work, but to your office, lab, or site. You'll enjoy each day more than the day before. The basic elements of your overall development should focus on lifelong education; engineering with a strong responsibility for positive societal impact; and globalism in attitude, knowledge, and responsibility, sprinkled with a good amount of balance of what life has to offer. All that, well stirred and brought to a boil during your career, will offer you a rewarding, enjoyable, and satisfying life and career.

Most of all, realize that you are not a person alone, but rather part of a worldwide team working toward a common goal of making the world a better place to live for all of humanity. May you become all you are capable of being and may your global experience serve you well in achieving your life's goal.

It is important to have trust, respect, and confidence in those you work with or for, as well as in those who you hire. Seek out those who provide that to you in reciprocation. Also remember that in any culture, TRC takes time to develop, and little replaces time in the long run—not even money.

Have pride in your past and faith in your future. However, don't live in the past or adopt hope as a singular strategy to assure that future.

Recognize that new initiatives are invariably developed by a champion or group of champions. However, once established, it is critical to ensure the sustainability of the program by cultivating future leadership. Sustainability needs to be defined in terms of not only resources but people.

Finally, to those who still may question why a global perspective is necessary as a part of an engineering education, I offer this:

- ❖ It's because every leader in history always recognized that the future of the country lies with the proper education of its youth and that education today mandates a global perspective in the global village of today and tomorrow.
- ❖ It's because 96% of humanity live outside the United States, only 20% of our citizens have passports, and less than 1.5% of our students participate in an international experience.
- ❖ It's because there exists a fundamental interdependence of all of our countries sharing the same planet including its energy, environmental resources, air, and water. Because a catastrophe for one of us becomes a catastrophe for all of us, and because global issues call for global solutions.
- ❖ It's because developed countries have an obligation and responsibility to those less developed, because famine of one country's children is famine for all of our children, and because we still need to learn to prize long-term cooperation more than winning a short-term conflict.
- ❖ It's because simply: it's the right thing to do. So just do it.

Chapter Summary: Takeoff Tips

- ❖ Global attitude and experience will benefit you and your engineering career.
- ❖ Why is global attitude necessary?
  - 96% of the world's population lies outside the United States of America.
  - 1.5% of domestic students have an international educational experience.
  - Our individual isolationism is in conflict with the global village reality.
  - Earth is a technologically borderless planet with increasing change rate.
  - Environment and energy are key global concerns: what are the technical solutions?
  - New norm: multinational corporations with employees globally located.
  - There is global economic interdependence among all countries.
- ❖ How to develop global competence:
  - Plan to create your opportunities for global understanding.
  - Do international travel as much and as early as possible.
  - Learn a second language.
  - Read all you can about other countries.
  - Study abroad for one or two semesters in your junior year.
  - Do an international internship for a semester.

❖ Major influencing factors:
  - Be self-motivated and passionate about your global objectives.
  - Pursue lifelong learning through formal education and experience.
  - Be both an engineer and a humanist: can it be done? How? What's the impact?
  - Use each driving force: need, change, imagination, and passion.
  - Think of leadership and teamwork as synonymous.
  - Never underestimate your abilities.
  - Pursue life with a passion in all respects.

Exercises

1. Why is it necessary to have a global attitude and global experience?
2. Why has your environment become everyone's environment?
3. How do you plan for global understanding through a planned experience?
4. Why and how should you study abroad for a reasonable period?
5. How do you decide on an international school? What are the considerations?
6. How would you plan for an international internship?

Summary Review of Parts 1 and 2 and Looking Ahead to Part 3

In Part 1 of *Ready for Takeoff!* you learned pre-employment success subjects, specifically, step-by-step strategies guiding you to get the right job, based on your skills, interests, values, and objectives. This included your career development actions as an engineering student: self-analysis, market research, writing your resume, use of your career center, and your successful job search campaign. The successful interview enables you to close the deal on a best job and prepare to start employment—aiming to achieve your unique career potential.

In Part 2 you have learned engineering functions and opportunities, explained by experts. This included an overview of industry by CEOs and engineering directors, project management, value engineering, quality engineering, lean enterprise, engineering professionalism, entrepreneurism, academic careers and graduate school, and becoming a global citizen. All the subjects in Part 2 are extremely important for a professional engineer in industry, government, or academia. Knowledge of these subjects as you enter your profession will give you a competitive head start to your career as a co-op student or new graduate.

Next, in Part 3 we will complete the total engineering career launch information package by focusing on personal and professional success skills, the importance of which was introduced by our engineering executives in Chapter 11. These subjects include self-reliance, planning and time management, empowerment and motivation, communication/interpersonal skills, teamwork, leadership, effective writing and public speaking, and transition to industry.

Evidence of the significance of Part 3 subjects was provided in Figure 2–1, page 17. As a reminder, executives throughout the United States were asked to rate new engineers on their preparedness for practice in eight areas, including teamwork, system design, leadership, integrative thinking, social/ethics/environment, math and science, market environment, and social sciences. The conclusion was that the average engineering graduate was underprepared in leadership, teamwork, and other soft skills. Only in math and science did the average undergraduate have preparation that exceeded what employers perceived as high value. Again, to quote the conclusion of the major research findings described on page 17 of Chapter 2: "With the exception of 'Math and Science,' there appears to be a wide discrepancy between the value expectations of the employer and the extent to which their employees are seen to be well prepared."

We refer to the study again only to reinforce the impact of the subjects throughout this text, including Part 3. Some of the material in Part 3 might appear to be common sense that could be taken lightly. But absorb these chapters and, once again, you will be off to a faster and better start than those engineering graduates who were strictly technically focused and not so broadly educated.

Note: With technology, the real test is your ability to apply theory toward useful results. So it is with the material in the Part 3 chapters: the concepts are easily understood. Here, the test will be for you, an engineering student, to recognize the importance of these non-technical subjects and live them in your personal and professional habits, which all employers seek.

Personal and Professional Success Skills

PART 3

CHAPTER 20

Self-Reliance, Planning, and Time Management

"Common sense is not so common."

VOLTAIRE (1694–1778)

Self-reliance, planning, and time management are parts of a fundamental skill package for a competent engineer, indeed any professional adult, to be effective. They're a starting point for you as a professional engineer, since you need to have your own personal act together before assuming leadership or teamwork responsibilities with others, including customers. It's critically important that you display a high level of personal responsibility as you take on project assignments. In fact, these skills are all about you landing the right job in the first place.

SELF-RELIANCE

It means just that—you can take care of yourself. It's great that you're studying engineering—a curriculum that is extremely worthwhile and more difficult than most people can handle. By definition, as an engineering student, you have above-average intelligence and motivation. Rather than encourage you to lay back with that comforting thought, however, I'd like to throw out a challenging question. Are you reaching your self-reliance potential? Are you what Stephen Covey calls proactive?

I highly recommend Stephen Covey's best-selling book, *The Seven Habits of Highly Effective People*, as an explanation of habits that you can learn in order to be successful, both personally and with other people. Tip: this book is extremely popular in the corporate world. In his book, Dr. Covey describes a human "maturity continuum" that goes from dependent (starting at infancy and being totally reliant on others), to independent (being able to take care of ourselves), and finally to interdependent (in which the independent person is able to effectively work with others, be a team member and leader).

"Be proactive" is Dr. Covey's Habit Number 1. He contrasts being proactive to being reactive. Reactive means that you respond to situations, either positive or negative, without exercising the freedom to choose your response. If somebody criticizes you, if the weather or the economy is bad, you let it get you down. Conversely, if something positive happens, you feel good. It's natural to feel positive or negative emotions under certain circumstances, but the key to being proactive and self-reliant is understanding that you have the freedom to choose. The freedom to choose begins in your brain when you embrace the commitment that you do have the power to elect your response to situations. You are not hot-wired from a positive or negative stimulus to an automatic, unthinking response. Between the stimulus and your response is a zone of freedom, and that choice lies within your brain. Dr. Covey points out that truly proactive people can, through self-discipline, program their brains to understand that they can choose the response to whatever confronts them. Your controlled thoughts turn into your controlled feelings, then into your controlled response or initiated action. That's powerful. That's extremely important. It means you're in control. And you don't need to be a psychology major to understand this concept and embrace it as the foundation of self-reliance. Self-reliance in action:

- ❖ ***Choose Optimism.*** Yes, between 2008–2010 the economy has been poor—like continuously bad weather. If you had been in the job market during this time and were a reactive person, you might have been overwhelmed with pessimism. You might have believed it to have been impossible to get a good job, based on the economy: a plunging stock market, corporate layoffs, and non-stop gloom reported in newspapers and on television. In fact, if you felt hopeless about job prospects, you might well have given up on the whole business of trying to get a good job and thrown in the towel. On the other hand, what happens if you choose to be optimistic about your engineering skills, your passion to contribute those skills to a market niche where there's a need, and your unwavering confidence that you will succeed? Whatever happens to be going on in the economy at the time you would, could, and will win with that positive attitude!
- ❖ ***The Secret of Proactive Optimism.*** What's the secret of proactive optimism and how does it relate to self-reliance? The answer is simple to understand, but the execution of the answer requires strong willpower. Take the case of the optimistic job hunter in a bad economy that was described above. Let's say you choose to believe that despite the economy, there are still good jobs out there and, in fact, there is less competition for those jobs since many people are sitting on the sidelines waiting for job prospects to improve. You are operating within what Stephen Covey describes as your "circle of influence" (things you can take action on and do something about) as opposed to the pessimist's "circle of concern" (the bad news about which you can do nothing but drag yourself down). So, the secret of proactive optimism related to self-reliance is that

if you choose to believe in a positive outcome and take persistent action based on that belief, you're likely to succeed.

An old sports adage goes, "Whether you think you're going to win or lose—in either case you're right." You're in charge of your attitude! Use it positively for your success and happiness!

You might say: "I choose to have a positive, optimistic, winning attitude. Any specific suggestions as to how to get there?" Here's how:

- ***Read This Book.*** Read it from cover to cover. Knowledge is power, and *Ready for Takeoff!* includes 32 experts who convey important information on everything from getting a job to being successful in your career. This information will supplement your technical engineering coursework and allow you to take self-reliant action to be a very competitive job candidate. You'll be highly attractive to employers.
- ***Work on Developing Your Self-Esteem.*** I'm not talking about egotism, arrogance, or narcissism—not at all; those qualities are offensive. The suggestion here is that you proactively decide that you are worthy of success. The effectiveness of your daily actions and the level you reach are either enhanced or limited by your self-confidence.
- ***Instill Desire.*** Are you looking for a good engineering job? How badly do you want it? The more passionately you want to get the right job for yourself, the harder you'll intelligently work, day after day, toward that goal. You'll be armed with resumes, business cards, and your market analysis of potential opportunities. You'll act on an ever-expanding network of people and companies you'll call on to connect your talents with their needs. You'll get invitations to discuss potential employment opportunities.
- ***Positive Self-Talk.*** We all have an inner voice (our voice) that talks to us on an ongoing basis. The major problem with that inner voice is that it's often too critical—"I messed up," "I'm disorganized," "I'm not smart enough," "That company would never hire me." What you need to do with that inner voice is hit the delete key in your brain when you're hearing thoughts that are unreasonably hard on you. You can choose to substitute that negative voice with positive personal growth affirmations that are constructive and don't tear you down. Certainly, acknowledge areas where you need to improve, but don't destroy your self-esteem by being your worst mental enemy. In fact, if necessary, you might borrow something like the self-pep talk quote by Stuart Smalley on *Saturday Night Live:* "I'm good enough, I'm smart enough, and doggone it, people like me!" Smalley did this as humor, but self-talk confidence-builders can be helpful. Make it genuine so that your brain believes that your inner voice is valid. Pumping yourself up is particularly important when you're heading into an interview with a desirable employer. In that case, mentally focus on reviewing your specific positive qualifications for that employer's particular needs. Review how you're going to be a STAR, per Chapter 8, page 78, on behavioral interviewing. Prepare to describe situations, tasks, actions, and results. Concentrate on the evidence of your achievements. Put

positive enthusiasm into your inner voice that you will carry into the interview when you are actually speaking.

- ❖ ***Positive Visualization.*** If you fear an upcoming interview and feel unprepared, you'll be less likely to do as well as if you're confident and well prepared. This is another of those observations that might appear so obvious that it's not worth mentioning, but at times obvious things might not get the attention they deserve. Most candidates going into an interview don't really focus on positive mental preparation. Better they should visualize themselves confidently and effectively discussing their specific attributes that will meet an employer's need. Do this: vividly picture yourself having a very positive discussion with the prospective employer who will be interested in you and what you have to offer. It'll pay off. Your positive visualization of the interview and any other challenging experience, including a public speech, is likely to carry over to your actual performance and end up as a winning, self-fulfilling prophecy. Successful people do this.
- ❖ ***Accept Negatives that You Can't Control.*** Short statement: if there is a negative situation that is beyond your control, accept it. It is what it is. Look past it and focus on something positive that you can control. Dwelling on uncontrollable negatives just drags you down and works against your best performance.
- ❖ ***Associate with Positive People.*** Birds of a feather do flock together. Positive people are generally happy and successful. Hang out with them—your and everyone's positive energy will rub off on each other for mutual benefit. They are likely to be the leaders of the organization, exchanging ideas to move the group ahead and enhance their own success as well. That's a success network you want to belong to. On the other hand, in many organizations there is a like-minded group of pessimists. They also hang out together and have "whine and jeez" parties; they say, "Aw jeez . . ." about almost anything. These are, quite frankly, the losers in the organization who habitually moan and groan about who got an undeserved promotion and why the company is treating them so poorly. Be polite to everyone, but definitely avoid hanging out with the pessimists.

Step by step, if you adopt the above favorable actions, your confidence and self-esteem will grow positively in such a way that you will feel worthy of success. It's your inner strength that will lead you to be a proactive, self-reliant person. Your determined confidence will lead you to believing in yourself. By developing your self-reliance, you're giving yourself a wonderful gift that you'll share with the organization for everyone's benefit.

SELF-RELIANCE POEM #1

If you think you are beaten, you are.
If you think you dare not, you don't.
If you'd like to win but think you can't,
It's almost certain you won't.
Life's battles don't always go

To the stronger woman or man,
But sooner or later, those who win
Are those who think they can.

W. D. Wintle

SELF-RELIANCE POEM #2

I bargained with life for a penny and life would pay no more,
However at the end of a long hard day I counted my scanty score.
For life is a just employer—it pays you what you ask,
But once you have set the bargain then you must bear the task.
I worked for a menial's hire—only to learn, dismayed,
That any wage had I asked of life, life would willingly paid.

Jessie B. Rittenhouse

- Now, hopefully, you've grasped the reality that you can choose your outlook. In doing so, you have unleashed tremendous potential to create your own future. Rather than waiting to get lucky and be handed a job offer or be directed what to do, you can visualize your future and make it happen on a daily and long-range basis.
- Based on this understanding you can literally create your own future. This self-reliance will go hand in hand with everything you have done in Part 1 of this book, that is, your self-analysis and well-directed career search toward the market niche that will best use your skills, interests, values, and goals.
- This can lead to a self-reliant journey of lifelong personal and career fulfillment. You will know where you're going and why you're going there. That conviction will enhance the likelihood that you'll end up successfully where you want to go. What a huge plus for your fulfillment and happiness! You can be an example-setting leader of a well-rounded, successful life. What a positive, meaningful future to look forward to!

PLANNING

Planning is what a self-reliant, proactive person does to achieve your objectives. It means writing the action steps you're going to take in order to achieve your goal. Planning is something that seems obvious and logical, yet many people don't do it. Why?

Reasons Why Some People Don't Plan

- They don't know what they want. Unless you know what you want, why plan?
- They are reactive, "going with the flow." They're waiting for someone else to do the planning and tell them what to do. So they don't really have a proactive goal.

- ❖ They don't want to take the time to plan. They just don't sit down and decide where they're going and the steps required to get there. They don't understand that they are going to spend more time resolving confusion than if they started with a logical, sequential plan. Not enough time to plan, but always enough time to correct mistakes.
- ❖ They can't predict the future. These people don't plan because they think that something will come up to disrupt any written plan. So they avoid planning and reduce their chances of reaching a goal.
- ❖ They fear not meeting a written goal. These people are afraid to commit to themselves. They don't want to stick their necks out and make a commitment for fear of embarrassment if they fail. Too bad they don't realize that making this commitment in writing is a huge motivator toward meeting a goal.
- ❖ They have an overconfidence of what they can carry in their heads. These people figure that they'll be able to remember anything that would go into a written plan. If the goal involves any level of complexity, they're wrong and will regret their overconfidence.

There are benefits of planning—choose which is best for you; the above reasons for not planning or the following benefits of having a plan.

Benefits of Having a Written Plan

- ❖ A written plan says that you have a goal.
- ❖ A written plan clarifies what needs to be done toward your goal.
- ❖ A written plan allows you to prioritize. It allows you to recognize what's most important and the proper sequence of action steps toward your goal.
- ❖ A written plan keeps you on course. Deviation from the plan allows you to be alerted to take corrective action to put you back on track.
- ❖ A written plan is the foundation of time management toward your goal.
- ❖ A written plan sets growth expectations and allows you to personally stretch.

Evidence of the Benefits of Planning —A Harvard University Study

Mark H. McCormack gives evidence of planning benefits in his book, *What They Don't Teach You at Harvard Business School.* He tells of a study of 1979 Harvard MBA students where they were asked, "Have you set clear, written goals for your future and made plans to accomplish them?" While 3% of the MBA students had written goals and plans, 13% had goals, but not in writing. However, 84% had no specific goals. Then he relates that 10 years later, the members of the 1979 Harvard MBA class were interviewed again. Here were the results: the 13% of the class who had unwritten goals were earning twice as much as the 84% who had no specific goals. What about the 3% who had written goals and plans to accomplish them? They were earning 10 times as much as the entire other 97%!

Evidence Sells. Our chapters on creating a resume and being interviewed emphasized that evidence sells; that is, concrete evidence of your ability sells an employer to give you a job offer. In this case, the Harvard MBA study of the benefits of a written life plan for the future is powerful evidence of what you, as a proactive person, might want to do.

The **action plan** is a fundamental business requirement. In Chapter 10, we discussed that as a co-op student or new engineering graduate, you'll meet with your boss during the first few days to determine what you'll do on your assignment. Your university or your employer may have a format for an action plan that you'll use to accomplish your expected results. In any case, you'll want such a plan for all the benefits described above. Figure 20–1 provides you with an example of an action plan model that you'll fill in, beginning with the "Deficiency or Need" (why you were hired) and ending with the action steps, accountability, and completion dates for your assignment. Strong advice: don't begin an assignment until you have understanding and acceptance with your supervisor as to what you're going to do and how you're going to do it, along with others who will also be responsible. This written plan will guide you, including clarification of who does what and when, from your first assignment on through your career.

ACTION PLAN

Deficiency or Need:

Specific Objective:

Results Expected: Measures

1.
2.
3.

Action Steps:	**Person Accountable**	**Due Date**
1.		
2.		
3.		
4.		
5.		

FIGURE 20–1 Action Plan

TIME MANAGEMENT

Time management is the final action step of this chapter. Proactive self-reliance and planning are the necessary prerequisites for you to take action toward your goals. Now it's time to put your self-reliance and planning skills into action.

Your values are the foundation of time management. Time management begins with knowing your values. Dr. M. Scott Peck MD, author of *The Road Less Traveled*, says it well, "Until you value yourself, you will not value your time. Until you value your time, you will not do anything with it." Ralph Waldo Emerson adds, "Nothing gives so much direction to a person's life as a sound set of principles." Therefore, before proceeding to the logistics of time management, you need to determine your fundamental values. These are your foundation principles upon which you build your personal and professional life. Take the time to list your most important values before we proceed to the details of time management. Some examples of fundamental values are: I am physically fit. I have high integrity. I lead a balanced personal/professional life. I value an engineering career. I am self-reliant. I am optimistic. I seek to fulfill my personal and professional potential.

My fundamental values are:

❖ ______________________________

❖ ______________________________

❖ ______________________________

My long-range goals that will fulfill my personal and professional life are:

❖ ______________________________

❖ ______________________________

❖ ______________________________

As we have mentioned previously, many, perhaps even most, people go through life with general thoughts of future happiness, but no specific goals or plans to achieve them. There is ample evidence, including the Harvard MBA study referred to in Mr. McCormack's book, that written goals give enormously increased probability that your dreams will be realized.

Hyrum W. Smith wrote a book on time management that I highly recommend: *The 10 Natural Laws of Successful Time and Life Management.* Mr. Smith originated the Franklin Planner, a time management system that has been of great help to me for years. Mr. Smith has since joined Stephen Covey to form the Franklin Covey Corporation offering time management systems (www.franklinplanner.com; click on Get Organized). In his *10 Natural Laws*

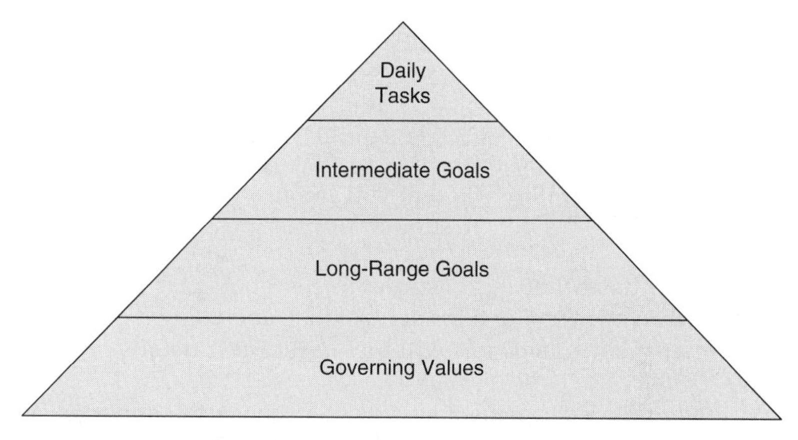

FIGURE 20–2 Productivity Pyramid From Smith, Hyrum W. *The 10 Natural Laws of Successful Time and Life Management.* Warner Books, 1994.

book, Mr. Smith diagrams a productivity pyramid, beginning its base with governing values, then in ascending order, long-range goals, intermediate goals, and daily tasks; see Figure 20–2.

Mr. Smith explains the choice of a pyramid to illustrate that in time management, the base has to be your governing values upon which your long-range goals are founded. As you move up the pyramid, your goals get more specifically focused so that at the apex, you are efficiently working on your daily activities that are today's best focus toward your long-range goals. Today is your productivity point. You plan for tomorrow but today is when you put your plan into action!

James W. Newman's book *Release Your Brakes!* describes that on a daily basis, we all tend to gravitate to a comfort zone—comfortable habits that we're used to. When we're in the comfort zone, we are coasting into what we've always done, requiring no significant effort. Reaching a goal means exerting the effort to leave a comfort zone. Therefore, comfort zones are tough to leave. You'll need to exert the energy to leave your comfort zone if you're going to grow and reach your goals. Accept the reality that it may be uncomfortable at first, but the best lies ahead.

There are a number of time management tools available, depending on your choice of electronic or paper systems. The important thing with time management is having a system that is easy for you to use, that is based upon your fundamental values, and that leads to your best use of each day. Whatever your system, Mr. Smith suggests the following rules.

Rules for Using a Time Management Tool Effectively

- ❖ Keep your day planner with you so you can refer to it for your prioritized tasks and appointments and make notes in it as necessary.

- ❖ Use only one calendar so that you don't waste time coordinating multiple calendars. Use this calendar to note all events and action required on a certain date.
- ❖ Commit to planning every day and make it a habit first thing in the morning or at the end of the day so that you regularly use your time management tool. Why? So that you can use your time each day most efficiently and effectively in support of your goals.

Mr. Smith has other suggestions in his book that include using a good reference system, using a master task list, and using a monthly index. Most time management systems will include specific details of how to use that tool.

Whatever time management system you choose, the important thing is that you do have a system that allows each day to be spent most productively toward what you want to do in life, so that each day you have your act together. This is all about self-reliance, it's all about planning. I suggest the acronym WIN = What's Important Now?

Mr. Smith suggests that we are all creatures of habits that can make us or break us. As a proactive person, you should believe that. You can choose your habits.

HABIT

I am your constant companion.
I am your greatest helper or your heaviest burden.
I will push you onward or drag you down to failure.
I am completely at your command.
Half the things you do, you might just as well turn over to me,
And I will be able to do them quickly and correctly.
I am easily managed; you must merely be firm with me.
Show me exactly how you want something done,
And after a few lessons I will do it automatically.
I am the servant of all great men and, alas, of all failures as well.
Those who are great, I have made great.
Those who are failures, I have made failures.
I am not a machine, though I work with all the precision of a machine,
Plus the intelligence of a man.
You may run me for profit, or run me for ruin;
It makes no difference to me.
Take me, train me, be firm with me and I will put the world at your feet.
Be easy with me, and I will destroy you.
Who am I?
I am HABIT!

—AUTHOR UNKNOWN

Do I need to say more? You choose.

Chapter Summary: Takeoff Tips

Self-Reliance

- ❖ You can take care of yourself. Are you reaching your potential?
- ❖ You have the freedom to plan your future.
- ❖ You have the freedom to choose your response to situations.
- ❖ Why not choose optimism and passion toward your life and career?
- ❖ Instill desire and self-esteem through:
  - Positive self-talk
  - Positive visualization
  - Accepting uncontrollable negatives
  - Associating with positive people

Planning

- ❖ Reasons why some people don't plan:
  - They don't know what they want and have no objectives.
  - They are waiting for someone else to tell them what to do.
  - They don't want to take the time to plan.
  - They think they can't predict the future.
  - They fear not meeting a written goal.
- ❖ Benefits of planning and having a written plan:
  - It says that you have a goal.
  - It clarifies what needs to be done toward your goal.
  - It allows you to prioritize and keep you on course.
  - It is the foundation of time management toward your goal.
  - It sets growth expectations and allows you to personally stretch.
- ❖ Evidence of the benefits of planning—a Harvard University study:
  - 3% of MBA students had written goals and plans. Ten years later, these students were earning 10 times as much as the entire other 97%.
  - 13% had goals, but not in writing. Ten years later, these students were earning twice as much as the 84% who had no specific goals.
  - 84% had no specific goals.
- ❖ The action plan is a fundamental business requirement list:
  - deficiency or need
  - specific objective
  - results expected: list the measures
  - action steps: list them along with the person accountable and the due date

Time Management

- ❖ Your values are the foundation of time management.
- ❖ List your fundamental values.
- ❖ List your written goals and plans.
- ❖ Reaching a goal means exerting the effort to leave a comfort zone.

- Choose a time management system and use it toward your goals:
    - Keep your time management tool with you.
    - Use only one calendar.
    - Commit to planning every day.
    - WIN = What's Important Now? That's for you to decide.

Exercises

1. Define self-reliance and self-reliance in action.
2. What are the benefits of self-reliance?
3. How can you become a self-reliant optimist?
4. What are the reasons why some people don't plan?
5. What are the benefits of planning?
6. Make an action plan for one of your specific objectives.
7. Why are your values the foundation of time management?
8. Give examples of your fundamental values.
9. Select and use a time management tool, if you haven't already done so.

CHAPTER 21

Empowerment and Motivation

In Chapter 20 we covered the subjects of self-reliance, planning, and time management. Now, Mr. Michael Rizzolo will build on these subjects.

MICHAEL RIZZOLO

Michael Rizzolo is President of Communications Services, Inc. He has been extremely successful in presenting seminars to employees of corporations, universities, and government organizations. Why are these organizations so anxious to have Mr. Rizzolo talk with their employees? Because he explains fundamental concepts of human behavior and motivation that are very important but not often used. He invites listeners to reach their full potential, both personally and on the job. That's good for business and it's good for you. Several points before we begin this chapter:

- Michael Rizzolo is not a magician; he's not able to get inside anyone's heads and motivate them on an ongoing basis. Only you can open the door to your own motivation and your pursuit of all that you can be.
- A chapter such as this does not appear in standard engineering textbooks. This will be a trip into the right hemisphere of your brain, a study of emotions and what they can do for you and others. As a total engineering career launch book, this subject needs to be covered. Understanding and acting on these right-brain concepts can be to your great advantage as a supplement to your technical skills.
- In reality, there is a normal curve in the engineering population—most of my students think Michael Rizzolo's information is great. You may or may not; you be the judge and use his message to the extent that it's helpful for you. I'll just add that successful people are big-time advocates of the continuous development of themselves and those around them. They use these concepts. Mr. Rizzolo is a very powerful and empowering speaker. I encourage you to absorb his thoughts on these printed pages

while visualizing him speaking with you personally; visualize a live performance where he invites you to be whatever you choose to be.

EMPOWERMENT OF YOURSELF

This is all about making the most of your potential in your personal and business life.

- ❖ There are many strategies to get where you want to be.
- ❖ It's all about influencing yourself. It's your choice.
  - Some will get started in their 20s toward reaching their peak.
  - Others are late bloomers and will start later.
  - Still others will never really start.

WHAT SUCCESSFUL ENGINEERS KNOW

The most successful people are those who demonstrate the ability to emotionally influence themselves and others to take action.

- ❖ You, as an engineering student, have tremendous potential. You are launching into a profession of incredible opportunity.
- ❖ You have the ability to influence yourself and others. The question is: are you doing it?
- ❖ Is it more difficult to get yourself to take action or to get others to take action?
  - Generally, the answer is that it's more difficult to get yourself going.
  - When you get yourself to take action, you can get others to take action.

FOCUS

Whatever we focus on, we get.

- ❖ We can determine our focus every minute and optimize it toward what we want. Break out of unproductive habits and focus on those things that really matter to us.
- ❖ Those people who can master their positive focus are the most successful.
- ❖ Contrast that with the mind-set of people who come to work with an eye toward going home. In fact, they literally "come to work to go home." They're just going through the motions without much emotion.
  - They watch the clock, waiting for 5 p.m. each day.
  - They focus on making it to Friday night and the weekend.
  - They live for their vacation each year.
  - They spend their life looking forward to retirement.
  - Have they enjoyed their careers? Have they enjoyed life?
- ❖ With your engineering degree, you have countless opportunities.
  - Within your potential, you can do anything you put your mind to.
  - You go to school to get the knowledge so you can create value in the marketplace. It's not what you know; it's what to do with what you

know that counts. You can be a technical specialist, a manager, an entrepreneur, or anything that suits your abilities and interests. Lots of options.
- With your good attitude, focus, and skills, you can control your destiny.
- Never let go of the things you love to do.

MANAGE YOUR DESTINY

You can manage your destiny starting now.

- ❖ ***Life Accumulates.*** The decisions you make now will impact your future.
- ❖ ***Listen to Your Brain and Your Body.*** What do you want to do with your life?
- ❖ ***Maximize Your Positive Experiences.*** There are many opportunities that you are now taking advantage of or that are just waiting for you to take action.
  - Get to know professors and others who can help your life and career.
  - Join professional societies.
  - Get relevant experience in line with your goals, skills, and interests.
  - Get the most out of these experiences. Focus. Grow. Enjoy.
- ❖ You also have the option of goofing off. That's part of our free society. It's your choice whether or not you want to be successful.
- ❖ If you continue to do what you've always done, you will continue to get what you've always gotten. If you have good habits that have brought you continued growth, success, and happiness, that's good news and you'll get more. On the other hand, if your habits have left you with pain and disappointment, you have an opportunity to change and improve your outcomes. You choose your habits and, therefore, your results.
- ❖ Albert Einstein reportedly said that one definition of insanity is doing the same thing over and over again and expecting a different result.

THE TWIN TARGETS OF HUMAN BEHAVIOR

The twin targets of human behavior are avoiding pain and moving toward pleasure.

- ❖ People will do way more to avoid pain than they ever will to gain pleasure.
- ❖ Don't criticize others or yourself. That's pain. That tears yourself and others down and undermines success.
- ❖ Our daily activities and our destiny are determined with what we associate pain and pleasure. The definitions of pain and pleasure are very individually determined. For example, some people get a pleasurable thrill out of seeing a horror movie that gives them an adrenaline rush. For others, it's painful to have the wits scared out of them.
- ❖ People spend a lot of money to have their emotional state changed away from pain and toward pleasure. If you have a skill, a product, or service

that will produce a positive emotional state change, you have a great opportunity for business success.

- ❖ If you are going to be influential, you need to understand your power to change your state of mind and emotion or those of others.

EMOTIONAL STATES

There are two ways to change a person's emotional state.

1. ***Change Your or Their Physiology—MOVE.*** If you move your body, you will change your state; call it e-motion. That's right; your emotion is connected to motion. People change their state when they get up to dance, engage in a sport, or walk around. The body-mind connection is powerful. Use it positively to change your state or someone else's state.
2. ***Change Your or Their Focus.*** Everyone asks themselves questions. What does this mean? What should I do about it? If you ask yourself a negative question that depresses you, you'll lose ground as a result of that question. Better to ask yourself a quality question. Quality questions are empowering and invite a positive outcome. The quality of the questions that you ask yourself will determine your outlook. Your outlook will determine your actions. Your actions will determine your destiny. You can ask questions of yourself and to others that are empowering and make tremendous positive differences in your life and those around you.

Suggested Action

- ❖ ***For Yourself.*** You have the power of positive self-talk, visualization, and movement. Positive self-talk includes empowering questions, visualization, and action. These options can move your emotion away from pain and toward pleasure with the result of being happier and more productive. You can change your physiology—you can move after a large meal that is making you sleepy. You can adjust your posture and body language to a more energetic level. You can slouch with your eyes on the floor while looking and feeling depressed. You also have the option of standing up straight, throwing your shoulders back, smiling, and being really positive and confident—maybe even pretending at first. The fact is, changing your visualization and body language can change reality. Acting more positive, confident, and happy can actually move you into that emotion. In short, you have the power to change your state. That may sound like B.S., and it is—if you define B.S. as your "belief system." Happy and successful people have positive belief systems. You either have a positive belief system or you can acquire one and make it continually better. Don't underestimate your potential for that development of your physical, mental, and emotional fitness.

❖ *For Others.* You have the power of infectious enthusiasm and asking empowering questions to help move someone who is in a negative state toward one that is positive.

- Let's use an example that's especially good for you and another significant person—the interviewer. Say you're going to an interview at the end of the day and you can see that the interviewer is tired, what can you do? For your sake and that of the interviewer, your job as a state changer is to positively influence the interviewer with your energy and enthusiasm to move from tired indifference into your territory of positive excitement about being a hot prospect for their company.
  1. One example of how you can change an interviewer's state is being able to enthusiastically describe your qualifications and how they relate to the company's specific needs.
  2. How about having some good questions about the employer that show your intelligence and keen interest in the company? Good questions are great state changers for a previously tired interviewer.
  3. One possibility is to sincerely ask the interviewer why he or she likes the company. That could change the state of the interviewer away from tiredness and toward enthusiasm. Then, mirror (respond positively to) that enthusiasm.

  Interviewers are always looking for choice candidates and they really don't want to be bored. So, know that you have the potential to move interviewers and other people into positive territory by changing their emotional states. When people associate pleasure with you, you change their state and they want to be in a relationship with you. The positive outcome with the interviewer can be a site visit or a job offer for you. That's a winner for you, the interviewer, and the employer.
- People spend a lot of money to have their states changed. For example: As an engineer, understand that consumers make their decisions for emotional reasons and then justify them with logic.
  1. Volvo sells on the emotion of safety.
  2. Luxury brand automobiles sell on the emotion of status.
  3. SUVs sell on the emotion of being youthful and sporty.

  Whatever the product, those designing it need to understand this human desire of the consumer. They want you to buy their product to have your state changed away from pain and toward pleasure. The leaders in a successful company know the emotions of their market niche and direct their designs and marketing toward it. You can be one of those leaders.

EMOTIONAL STATES AND FOCUS

If you focus on what you don't want, you get more of it.

❖ When the coach says, "Don't drop the ball," dropping the ball is what's on your mind, and it is more likely to happen.

❖ The same is probably true when your mother said, "Don't spill your milk."

- ❖ When the judge says, "Strike that from the record," the jury is inclined to remember that point more than other testimony.
- ❖ An obstacle is something you focus on when you take your eye off the goal.
  - Deal with your obstacle as a goal and you can remove the obstacle.
  - For example, ask yourself, "How can I lose 10 pounds and enjoy the process?"
- ❖ You're always an influencer and always in sales.
  - You have to believe in yourself to sell yourself and the product.
  - As an engineer, that's your challenge: to sell yourself and your ideas.
- ❖ Nothing has meaning until I give it meaning. A group of people can experience the same situation, and each will interpret and remember it differently.
  - For you as a professional, this means keeping your eyes and ears open for what other people—that is, customers, teammates, and so on—regard as a turn on or turn off.
  - You may think an idea is great, so you give the idea a meaning of greatness. Others may not. If others don't like the idea, you have choices: get more information about why they don't like the idea, which may lead you to the same conclusion—or you can persist with the development/revision of your idea, recognizing that it will catch on with others as the idea turns into a productive reality.
- ❖ Make yourself continuously valuable on the job.
  - Employers look for those who want to work rather than have to work.
  - Define your success. If you're on the right job with the right attitude, you can have success every day.
  - Professionals put themselves in a positive state regularly. That includes you if you are going to be a successful student. It'll also determine your professional destiny. Make it a habit to put yourself in a positive state each day, and succeed with enjoyment. Your positive enthusiasm will be infectious and help others to succeed. In short, you'll be a leader whether or not you have a leadership title.
  - Those who always focus on negatives are less likely to be happy or successful. Why do they choose a miserable attitude? Most likely, they didn't consciously choose it. Most likely it's habit, without realizing they have an alternative choice of attitude.
  - Employers seek to hire people who create value moment after moment. A great attitude is the visible key to that continuous creation of value. Value creators with a great attitude get hired and promoted. They create value themselves and inspire others to create value.
- ❖ Hiring is based on accomplishments and attitude. Bring positive evidence of both into an interview and you're likely to be on the list of top candidates.
- ❖ There are no failures; there are only outcomes. That observation is attributed to Thomas Edison. He and Wilson Greatbatch are famous for not knowing

the meaning of the word "failure." If you have a negative outcome, you have the positive option of asking: What happened? What can I learn by that? Powerful questions, usually asked by successful people! A negative outcome can be a huge growth experience if you make it a positive learning experience. It's all in your choice of attitude and how you deal with everyday outcomes.

❖ Success equals being able to feel, on a regular basis, those emotions that are most important to you. Success is not an event or destination. Rather, it's a journey and the feelings that go with it. It's the opportunity to feel happy that you're doing something meaningful and fulfilling, by your own definition.
  - There are famous, wealthy, and powerful people who are unhappy. Are they successful?
  - There are people with very little money who are doing something that they really love and that gives their life great meaning. Are they successful?

OUTLOOK ON FAIRNESS

Life is not fair—perhaps it's not. In fact, who said that fairness in life is an entitlement? Michael Rizzolo says that some people who whine repeatedly about the unfairness of life deserve the definition of fair as: F.A.I.R. = Failure to Assume Individual Responsibility. Certainly there are unfortunate cases of health, accidents, and so on, but many people have become very successful despite hardships or misfortunes. Helen Keller, President F. D. Roosevelt, Christopher Reeve, and Michael J. Fox are examples. This and the previous chapter are all about self-reliance and doing the best you can with what you have. What else can anyone ask, along with being a law-abiding citizen?

THE SUCCESS FORMULA

Michael Rizzolo offers a success formula that he introduces with a powerful example. Roger Bannister was the first human to run one mile within four minutes. Throughout history it was regarded as physically impossible for a human being to run a mile in under four minutes. Mr. Bannister thought differently, and you can too in order to fulfill your dream. The success formula is:

1. *Know Your Outcome.* Roger knew his outcome. While cynics scoffed, Roger was committed to his idea of running a mile in under four minutes. You too can have an outcome that you're determined to achieve. Sometimes family and friends won't support you if you're aiming high; perhaps they're cautious and want to protect you; perhaps they're envious of your goal. But don't let anyone inappropriately stand in your way

or bust up your dream. Have a plan. A written plan. Commit yourself to your outcome.

2. ***Take Action.*** Roger ran. He trained. You, too, will achieve your outcome by taking action. Some people have an outcome in mind but don't do anything. The result? It's obvious. So, to achieve your outcome, get at it! Start moving with enthusiasm toward your dream!
3. ***Notice What You're Getting.*** Roger tried various approaches to improve his performance and enhance the possibility of reaching his goal. You too can measure your progress in your career and life. Are you improving and satisfied with your progress? You can't manage what you can't measure. You can measure quantitative things like your grades. You can also measure abstract things such as relationships. Taking stock once in a while is very important, whether it's in the quality of your professional or personal life. Are you progressing as you want to? What can you choose to do, with enthusiasm, to improve toward your goal?
4. ***Change Your Approach and "Go Until."*** Keep trying. Children will keep trying until they learn a skill. Adults are self-conscious and more easily discouraged if they don't accomplish something after the first few attempts. So, knowing this, keep at it until you reach your goal. Henry Ford was one of the first to say that "whether you believe you can or can't, you're right." Remind yourself of Thomas Edison's view that "there are no failures, only outcomes." If it is not the outcome you want, ask yourself, "What can I learn from that?" You can learn from numerous outcomes on the way to success, especially if you persist in your optimistic pursuit of your goal. Roger believed he could beat the four-minute mile. He used multiple approaches until he did it.

You need to have goals and take action. Goal-oriented action leads to results.

Use the success formula by knowing your outcome, taking action, measuring your results, noticing what you're getting, then changing your approach as appropriate, continuing to take action, and going until.

On May 6, 1954, Roger Bannister broke the four-minute mile using the principles of the success formula. For thousands of years, the four-minute mile was an impossible human achievement. Then, after May 1954, some people said, "If Roger can do it, I can do it." With that thought in mind, a number of people broke the four-minute mile in 1955. Now, Olympic track hopefuls from nearly any country need to break the four-minute mile just to get on the team.

As a personal take-away, realize that your brain can accept the belief that you will achieve goals that you want. Let your brain deeply absorb belief in your chosen outcomes. Then, use the success formula and other ideas in this chapter to reach your goals. And remember, success is not a destination; it's having the emotions that you want day after day as you work toward your goals. With that, best of success to you in realizing emotional happiness and being all that you want to be.

Chapter Summary: Takeoff Tips

- ❖ The most successful people can influence themselves and others to take action.
- ❖ Whatever we focus on we get. We can determine our focus toward our goals.
- ❖ You can manage your destiny, starting now. Maximize positive experiences:
  - Get to know professors and others who can help your life and career.
  - Join professional societies.
  - Get relevant experience in line with your goals, skills, and interests.
- ❖ The twin targets of human behavior: avoiding pain and moving toward pleasure:
  - People will do more to avoid pain than to gain pleasure.
  - Don't criticize yourself or others. That's pain, and it undermines success.
- ❖ Emotional states: there are two ways to change a person's state:
  - Change your or their physiology and move.
  - Change your or their focus. Quality questions invite a positive outcome.
- ❖ Consumers make decisions for emotional reasons and then justify them with logic:
  - As an engineer, you need to know this.
  - Successful companies know the emotions of their market niche and direct their designs and marketing toward it.
- ❖ Emotional states and focus:
  - If you focus on what you don't want, you get more of it.
  - An obstacle is what you focus on when you're distracted from the goal.
  - Deal with your obstacle as a goal and you can remove the obstacle.
  - You're always an influencer and always in sales.
    1. You have to believe in yourself to sell yourself and the product.
    2. As an engineer, that's your challenge: to sell yourself and your ideas.
  - Professionals put themselves in a positive state regularly.
- ❖ Make yourself continuously valuable on the job:
  - Employers look for those who want to work rather than have to work.
  - Being on the right job with the right attitude leads to success every day.
- ❖ The success formula:
  - Know your outcome.
  - Take action.
  - Notice what you're getting.
  - Change your approach and "go until."

Exercises

1. How can you manage your destiny, starting now?
2. Explain the significance of emotional states and focus.
3. Explain success relative to your regular emotional state.
4. What is the success formula? Give an example of how you use it.

CHAPTER 22

Interpersonal Skills

Are interpersonal skills important for an engineer? You bet! Obviously, you need to be a technically competent engineer when you graduate. But with your engineering education, you can do many things. Be assured that you'll be doing more than crunching numbers during your professional career. Whether as an individual contributor, manager, or project manager, you'll need to know how to communicate and deal effectively with people on the job in order to be successful. As a co-op student or new graduate, you may be on a design team, on a plant floor, on a construction site, with customers, or many other possible situations where dealing with people effectively will be crucially important.

During your career, your effective interpersonal skills—your people skills—will be as important as your technical skills. In fact, if you're promoted into a leadership position, your people skills will be a dominant qualification for success.

Let's bring interpersonal skills to life. How about getting some views from experienced engineers who can share first-hand experience about people issues in a technical organization? It's a privilege to have, as contributors, four engineers who will discuss dealing with people on the job in various functions, including testing, manufacturing, quality, and construction.

DANIELLE BUCHBINDER

Danielle Buchbinder is a test engineer with Dialogic Corporation, a former Division of Intel. She tests hardware and software products for the telecommunications industry. Ms. Buchbinder received a BS degree in Electrical Engineering and has 10 years of experience. Below is some advice from her.

❖ ***Live the Golden Rule.*** Everyone's heard it, but not everyone practices it. "Do unto others as you would have others do unto you." Or in more modern terminology, "Treat everyone like you want to be treated."

That's common sense, but it pays to remind yourself of that when you're dealing with people. It's a huge success principle.

- ***Be Friendly and Respect Everyone.*** Whether it's the president of your company or the person who empties your wastebasket, everyone is important and deserves your smile, a "good morning," and respect. Also, everyone has different skills and functions, all of which are vital to making the organization click.
  - ***Technicians.*** As an engineer, you're likely to be working with technicians who have less education than you. Never, ever regard them as "just a technician." You'll need them, big time. Especially as a new engineer, you'll find that they have considerably more knowledge and experience in their specialty than you do. Ms. Buchbinder was fortunate to work with a great technician who knew everything about the hardware being tested. They built an extremely valuable relationship of trust. From him, she learned how to learn and learned how to teach. She learned not to make mistakes.
  - ***Maintenance People.*** Under the heading of "everyone's important," what would happen if you treated the person emptying your wastebasket with disrespect? You guessed it: soon your wastebasket might remain untouched and overflowing. That actually happened when an engineer with a superior and unfriendly attitude was disrespectful to the person emptying wastebaskets. Don't let that happen to you.
  - ***Top Management.*** Job titles matter, but get to know the person behind the title. You can do that in a friendly way, without sucking up to anyone in the organization, including top management. Being a friendly, interested person is always appropriate. When appropriate, let your management get to know you and what you're doing. Express enthusiasm for the goals of the organization and you could discuss an appropriate subject of mutual interest. Earn their friendship and trust.
- ***Communicate Assertively When There's a Problem.*** In a problem situation affecting you, you don't want to be passive, nor do you want to be aggressively hostile. In real life, Ms. Buchbinder found it most effective to use a friendly, but assertive approach to honestly lay out a problem directly with another person. She started by listening to the other person and understanding their viewpoint. Then she explained how the problem was impacting her. From there, they were able to work out a solution that was mutually successful. Ms. Buchbinder could have ignored the problem or she could have gotten angry and turned it into a fight. Instead, by addressing the issue privately in a direct and friendly way, they were able to solve the problem and strengthen their professional relationship.
- ***Ask for Clarification of Your Assignment.*** Don't hesitate to ask if you don't know what's needed on a problem, if there's a request that needs to be made, or if there's a pertinent question you don't know the answer to. Don't guess on an important matter. Be open with your boss. Also, get to know the specialists who can answer your questions.

- ❖ ***Save Important E-Mails into Folders.*** That includes messages that you or someone else sends regarding project commitments. At a later time, you may find such messages to be invaluable evidence to avoid an argument and support your actions or requirements on a project.
- ❖ ***It's a Team.*** Specifically: **T**ogether, **E**veryone, **A**ccomplishes, **M**ore. This means everyone unselfishly working together to achieve the organization's goals. When you're part of the team, it's not about you—it's about the team goals and results. You may be working with some people you don't like, but focus on people's strengths that you can respect and focus on the goals. Get to know the people on your team and what motivates them. Ms. Buchbinder agrees with the well-known observation that "it's amazing what can be accomplished when it doesn't matter who gets the credit." Teamwork is the backbone of most organizations, and teaming involves interpersonal skills on everyone's part to achieve the desired goals. Chapter 23 is devoted solely to teamwork—it's really that important.
- ❖ ***Be Motivated.*** Ms. Buchbinder quotes Henry Ford, "It's been my observation that most people get ahead during the time that others waste." She agrees with Henry Ford. Do your best. Focus on the project. Share credit.
- ❖ ***Get Co-Op Experience.*** Finally, Ms. Buchbinder advises engineering students to get co-op experience. You learn people skills in the workplace. You also can explore your options, get applied technical experience, and set yourself up for a job offer when you graduate. She had a co-op assignment with Lockheed Martin Corp. and received a job offer from them when she graduated.

JOHN NIBLOCK

John Niblock is manager of Packaged Systems at Glauber Equipment Corporation, a manufacturer of skid mounted equipment for pumps and compressors for power generation, including gas turbines and alternative energy systems. Mr. Niblock holds a BS degree in Mechanical Engineering. He has experience as a plant engineer, project engineer, engineering manager, and project manager. Here are his views.

Working with People in a Manufacturing Environment

A manufacturing environment is a lot of fun. It's exciting, you get dirty, you wear jeans to work. It's a great leadership experience, making products and working with people. Here are some things to know.

- ❖ ***Respect.*** You've heard it from Ms. Buchbinder and you'll hear it from others: respect everyone and earn their respect in order to be effective and get things done. This is fundamental to all relationships. Listen to people, understand them. They will start to listen to you and respect your opinions. People appreciate respect and they appreciate sincere compliments.

❖ ***Leadership.*** The minute you come onto the shop floor as a plant engineer, you will hopefully be seen as a leader, since people will be looking to you for advice on how to do what you're asking them to do. So suddenly you're in a leadership position and to be an effective leader you'll have to give and get respect. That can be a challenge when you're in your early 20s leading some people who may be twice your age and have worked in that plant since before you were born. Will you be seen as a young whippersnapper or someone who's on the ball and deserves respect?

❖ ***Communication.*** Whether you are in a union or nonunion shop, communication will be one of your biggest challenges. Again, you need to respect people and listen to them, but if you're supervising and explaining something, you need to be a sales person. If you're explaining a new concept, you may need to use a classic teaching concept: tell them what you're going to tell them, tell them, and then tell them what you told them. Obviously, communication is two-way, and you'll want to check for understanding along with any questions. Confirm that they buy into the message—don't just assume it.

❖ ***Motivation.*** Individuals and groups differ. One group just wants to do the minimum; it's a leadership challenge to get them energized. Another group is motivated and wants the plant to do really well; this group comes in every day and they're on the move—looking for improvements and problems to solve. As a manufacturing plant engineer or supervisor, it's important to motivate positively and inspire everyone to pull together toward the same goals. Wherever possible, get your manufacturing team to realize the individual and group benefits of doing well. And when they do well, everyone likes a compliment—a pat on the back—"nice going; you've done a great job." Enthusiasm is contagious, especially if it's genuine. If you're a genuinely enthusiastic leader, you're much more likely to get support from your people.

❖ ***Teach.*** Teach others what they don't know. You're an engineer and can help them with your knowledge to solve manufacturing engineering problems.

❖ ***Learn.*** Learn from others what you don't know. Your machinists, welders, and technicians will be your teachers. You're going to watch what they do and you'll learn a lot. You'll learn specialized skills from people who've been developing these skills for many years. You'll learn what works, what's been tried before, and what has not worked.

❖ ***Use the Resources That Are Out There.*** There's talent on your manufacturing team. Use it. Maintenance. Technicians. Engineers. Don't feel that you need to know all the answers. There are other resources who can help you and with whom you can share knowledge and assistance.

❖ ***Dealing with People Outside of Manufacturing.*** Plan to do business with others outside of the immediate manufacturing environment. That could be design engineering, sales, purchasing, human resources, accounting, or any functional group within the company. You might be doing important business with an administrative assistant or the president of the company. You may be in contact with a customer or a vendor. In all of these cases, using the principles of good interpersonal skills will serve you well. Listen. Persuade. Collaborate. You'll get problems solved positively by using whatever is appropriate from the spectrum of interpersonal skills, all while enhancing your reputation. Tip: screaming and yelling is no longer in vogue! Using positive relationship building skills, you'll broaden your network of people you can help, and who'll help you in future situations. That's good for you today and is a great formula for career growth.

- ❖ ***Gain from Co-Op Experience.*** Mr. Niblock agrees with Ms. Buchbinder that co-op assignments are a great opportunity to learn. For example, find out about manufacturing. Based on that experience, would a job in manufacturing interest you after you graduate? The more engineering environments you're exposed to, the better your opportunity to decide on a career that combines your talents and interests. Determine that and it'll be a winner for both you and your future employer.

TANYA FARRELL

Tanya Farrell is quality leader at Continental Automotive, Inc., formerly a division of Motorola. Continental Automotive manufactures sensors and other electronic components, such as transmission control modules and emission controls for automobiles. Ms. Farrell has a BS degree in Electrical Engineering. She interacts with many people at Continental, including operators, engineers, and managers. An especially important job for her is communicating effectively with customers about new and current products.

Communication

Communicating is critically important. Ms. Farrell breaks down the subject of communication into three parts: **who** you are communicating with, **what** you are communicating, and **how** you are communicating. The following are her thoughts on each of these communication components:

Who you are communicating with?

- ❖ Is it a customer? Customers are the lifeblood of any business; they're buying our products, and those products need to be of top quality. Customers need to be satisfied. They're asking you for results. Listen carefully to their needs, including any suggestions or complaints. The listening part is super important so that you understand the issues before any action is taken. Then, take action promptly and effectively.
- ❖ Is it a supplier? Suppliers are giving us something that will help to complete the job. They're giving us the materials needed to produce top quality products. When communicating with a supplier, it's important to let your needs be known, including precise quality specifications and other requirements, including schedule.

- Is it internal communication within your organization? You need to communicate with your boss and deliver results that are expected. The same is true with other internal customers within your organization who you need to give something to. Or you may be an internal customer who needs to get something from someone. Again, you need to communicate and follow up with action to get necessary results.

What needs to be communicated? Ms. Farrell explains what needs to be communicated **to** you, and then **by** you.

- Communication **to** you
  - You need to know the big picture, including the company and departmental goals from your boss. Specifically, you need to know your boss's expectations of you. Before you start a project, it's important to know the performance criteria upon which you're going to be appraised. Salary increases are based on performance, therefore, it's good to know the expectations so you can meet or exceed them. So, communication to you from your boss is vital. Be sure there is mutual understanding.
  - As you join an organization, get to know others, besides your boss, who can help you succeed and from whom you can receive important communication. That includes team members and potential mentors who can coach you and accelerate your learning curve. Some of these people and company history resources can give you lessons learned from the past. This includes information about successful projects that may be an inspiration for your actions. Lessons learned also include failures and what to avoid.
  - Communication **to** you entails your responsibility to listen or read attentively so that you don't misunderstand. Ask questions or confirm the communication to you before you launch some important action. That way, you will be able to avoid the misfortunes of others who did not take this step.

- Communication **by** you. Ms. Farrell summarizes types of communications by you:
  - ***Results.*** Your technical and business results should be centered on facts not opinions. Numbers speak. Financial results are critically important toward profit.
  - ***Timing.*** There's always a schedule. If you can't meet a deadline, say so in advance and be up-front with a proposed solution.
  - ***Status.*** You should be prepared to report your project's status—i.e., is it completed? Does it require more work? Get agreement with your boss or customer on your plan of action toward completion. Projects need closure, and status reports as well as final closure are extremely important. Your boss needs confirmation that the project is completed.
  - ***Accomplishments.*** You'll want to let your boss know your specific accomplishments for the company. That's the basis for your performance appraisal, salary increases, and promotion opportunities.

- ***Constructive Feedback.*** This is very important for continuous improvement. Companies look for employees who suggest ideas for better performance.
- ***Be Approachable and Friendly.*** Ms. Farrell is very busy, but she feels that it's important to get out on the manufacturing line and get to know people while establishing friendly approachability. As a quality manager, this enables Ms. Farrell to have a close connection to the people and production processes that lead to high quality.
- ***Express Your Uncertainty with Confidence.*** Admit it when you don't know something such as when someone asks you a question or asks you to do something. Find out what you need to know, then check for understanding.

How should you communicate? Verbally and nonverbally. Ms. Farrell will explain both:

❖ Verbally

- ***With Clarity.*** Make clear points. Summarize. Don't give excessive detail.
- ***Call People by Name.*** That's a comfortable, friendly relationship builder.
- ***Meet People in Person.*** You can phone, but meeting an individual personally is key to building rapport and ensuring there's understanding about what needs to be done.
- ***Be Detailed or Concise.*** Know who you are talking to and have the flexibility to know whether detailed or concise communication is most appropriate for that person. It's also important to know when detailed versus concise communication is appropriate.

❖ Non-verbally

- ***With Diligence.*** Be professional. If you're late or sluggish, that communicates a lot, without speaking. First impressions are powerful—either positive or negative.
- ***Show Interest.*** People can see whether you're interested or not, and they respond accordingly.
- ***Smile.*** If you make it a habit to smile at people, it's a great way to show that you're friendly toward them.
- ***Project Reports.*** A well-written project report makes a very positive impression. Co-op students have impressed their management by preparing concise executive summaries of results: how much money or time can be saved, and so on.
- ***E-mail.*** It is a fast, efficient way to communicate factual information to a person or group. Be concise. Caution: an e-mailed message can be misunderstood, since it's not an immediate two-way communication. Be aware of e-mail etiquette, especially when to copy others on your message. Ms. Farrell suggests that on important or controversial matters

you should meet with or phone a key involved person, before the e-mail is sent in order to make sure there is positive understanding.

- ***Be Early.*** It shows self-discipline and respect.
- ***Show Flexibility.*** Adjust your style to who you're dealing with.
- ***Be Assertive.*** You should be assertive as opposed to passive or aggressive. Be honest and open.
- ***Lead a Meeting.*** You don't need to know it all; you just need to bring the appropriate people together. Leading a meeting is good experience and shows that you have initiative. It demonstrates that you can communicate, coordinate, and facilitate.
- ***Respect Coworkers and the Facility.*** Build a reputation for respecting both your fellow employees and the place where you work. As an example of respect for the facility, Ms. Farrell mentions a person who picks up a piece of paper from the floor rather than walking right by it; that's respect and gains respect from others.
- Read Stephen Covey's *The Seven Habits of Highly Effective People.* Ms. Farrell impresses an engineering class by reciting Stephen Covey's *Seven Habits* by memory—she emphasizes she didn't practice this beforehand. She stresses the importance of this book to engineering students for their personal and professional success. This has proven to be so in Ms. Farrell's case since she has lived the seven habits and has been very successful. She read the book as a student in my engineering career development course and amazingly, 13 years later, was able to spontaneously state the seven habits accurately and in order. John Lupienski, another of my students, said that he credits getting a job offer from Motorola when he mentioned that he had read Covey's *Seven Habits* book during a technical interview. I encourage you to read this and other self-development books as part of your professional and personal success journey. *The Seven Habits of Highly Effective People* is required reading for management in many companies. Read books like this and act on the contents; it'll immediately set you apart from other engineers who are focused only on their technical specialty. As a co-op student or graduating engineer, be technically excellent, but for maximum success, broaden yourself with skills described throughout *Ready for Takeoff!* and others in the recommended reading at the end of this book. Ms. Farrell is a living example of technical excellence combined with enthusiastic people and business skills. Be like Ms. Farrell along with our other contributors and write your own ticket!

Enthusiasm

Ms. Farrell closes by quoting Norman Vincent Peale, "Think enthusiastically about everything, but especially about your job. If you do, you'll put a touch of glory in your life. If you love your job with enthusiasm, you'll shake it to pieces. You'll love it into greatness. You'll upgrade it; you'll fill it with prestige and power." Signing off, Ms. Farrell tells engineering students, "Your attitude makes a world of difference. I enjoy my life and work every day!" Just like Ms. Farrell, you can choose a positive outlook that can make that world of difference!

STEVE PERRIGO

Steve Perrigo is a project manager with Turner Construction Company. Mr. Perrigo holds a BS degree in Civil Engineering. He has considerable experience dealing with people in the construction industry. Here is Mr. Perrigo's special insight about people in a construction environment and engineering in general:

Construction is a very exciting, demanding business, and it's all about communication. Every day, if you're not on the phone, you're talking with someone. If you're afraid of communication or afraid of commitment, construction is not the business for you.

Basics of Communication

- ❖ ***Listening and Speaking.*** Listen carefully, digest and analyze, then speak.
- ❖ ***Honesty and Fairness.*** Whatever you do, have high integrity at all times. If you start cutting corners on integrity and playing games, you'll be figured out quickly and will lose all your credibility.
- ❖ ***Respect.*** Everyone has mentioned it, and it can't be emphasized enough. Treat everyone with respect, and treat them like a peer. Make it a point to have valuable two-way communication.
- ❖ ***Say What You Mean and Mean What You Say.*** Do your fact-finding, do your research, then say what you're going to do and go do it. Get it done. That may mean a double (second) shift, but when you make a commitment, you have to deliver. Your reputation of keeping commitments will go a long way in your professional and personal network. When people know they can count on you, they'll be there to help you if you need it.

Diversity

Building on the basics of communication, a construction engineer—or an engineer in any environment—needs to understand the various types of people they are working with. There is a wide cross-section. On a construction job, workers can range in age from 18 to 65. You have some people on the job from beginning to end, and others doing a specialist job for a short time. Their personalities, tastes, and motivations cover a broad spectrum. You have to realize this when you're communicating with people, whether it's an individual or a group. Flexibility is required; you can't always be a driver pushing for more, and you can't always be laid-back. You need to adapt your communication approach to the situation; at various times, you need to be a listener, a decision maker, a leader, or a delegator. Now, let's look at some of these different types of people you will be leading.

Types of People on the Job

- ***The Planner.*** This person is organized. He knows what he's doing today and tomorrow as well as what's expected of him. You're lucky to get one planner on an entire construction job. Right now Mr. Perrigo is managing a large project that has between 400 and 500 people coming in and out of the job at any point in time. There's a consistent core of 40–50 workers. Of that, there's one person you can regard as an effective planner. You're lucky to have that person. Let him have freedom to plan, perform, and motivate others.
- ***Stuck in Their Ways.*** "I've been doing this job for 30 years and I don't care what you say—I'm going to do it my way!" How do you motivate a person like this, especially if their approach is ineffective? Maybe you work with the person and observe their approach. Then, in a friendly way, suggest something different. You will have to demonstrate that another technique is better. You can help get that person or group unstuck and moving ahead if you respect that them, demonstrate the benefit of an improved approach, and provide motivation to accept it. You also get an "A" in leadership.
- ***Mr. Negativity.*** "This job is bad news and everyone's out to get me." Every person and situation is different, but in this case, you could start with explaining the positive significance of the project. "We are building a school and that's important for the kids and the community"—whatever you think that person might value. You can change people's thinking if you show them the positives. Even someone performing a repetitive, narrow-scope job can be lifted if you help them see the big, positive picture. Maybe you heard of the mason who said, "I'm not just laying bricks, I'm building a cathedral." Finally, with Mr. Negativity, try to quickly clear up that person's paranoia or any actual problems with fellow workers. Or, maybe that person should find another employer; you don't have time or the qualifications for psychiatric counseling.
- ***The Survivor.*** "I'm on an island, being ignored. I'm on my own." Sometimes, a person doing field construction does need more support because of unexpected changes or other challenges. In this case, the construction leader will step in to keep the project going in the right direction and on schedule. That leadership can mean building rapport with the person and providing the support to get necessary things done—perhaps by getting the right blueprints or whatever is needed. Positive results happen when a person is informed, supported, and empowered to feel included as part of the team. That person is no longer in the dark.
- ***Chicken Little—The Sky is Falling.*** "Oh, my gosh, the blueprints are wrong! I've got four guys standing here, and I'm losing money. You've got to get this problem fixed!" Some people become frantic and lose their heads at a moment's notice. The project manager needs to remain calm when faced with this kind of "crisis" situation. The manager needs to

calmly sit down with Chicken Little, perhaps a plumbing subcontractor, agree on a definition of the problem, then the most logical steps for a solution. The immediate problem may take some time to solve. In that case, the manager might direct the subcontractor to a different part of the project where equally necessary plumbing work can be done with positive results, including nobody losing any money. The bottom line here is: keep calm and consider options when dealing with difficult people.

- ***I'm Only Going to Do What's Good for Me—Nothing Extra.*** In this case, we are talking about a number of subcontractors and people facing a large project—gutting and re-building a wing on a school in just two months. The scope of the project and limited time seemed overwhelming. The leadership challenge was for the project manager to get everyone to be confident and motivated that the job could be done well and on schedule. So, what to do? Mr. Perrigo, as project manager, met with each of the subcontractor supervisors and described the big picture, including the importance of the project and the necessity of the short time schedule. He emphasized the need for everyone to work together in order make the project succeed. Subcontractors were asked for their opinions as to how to proceed and were commended for the work they'd done in the past. In short, everyone was involved and empowered. The whole team became motivated and morale was high. The students were ready to return to their school by the end of August, but surprise! In just two months, the school had a new wing and 21 brand-new classrooms. Everyone involved was happy. That's an ideal ending to a major, challenging project. It wouldn't have happened without excellent leadership, communication, and teamwork.

So, did you enroll as an engineering student with the idea that you would be learning communication and people skills to this extent? More likely, you're an engineering student because you are interested in and good at math and science. An engineering curriculum in a university obviously and appropriately focuses on technical subjects. But, making a habit to practice the extra skills presented in this chapter will give you a significant advantage next to others who don't have this understanding of people and handling the kinds of situations that arise in business. Technical Skills + People Skills = Greater Success!

I conclude this chapter by strongly encouraging you to read another book, Dale Carnegie's *How to Win Friends and Influence People,* a timeless classic on people skills that was originally written in 1936. Dale Carnegie summarizes his Part 2 chapters:

Six Ways to Make People Like You

1. Become genuinely interested in other people.
2. Smile.
3. Remember that a person's name is the sweetest and most important sound in any language.
4. Be a good listener. Encourage others to talk about themselves.

5. Talk in terms of the other person's interests.
6. Make the other person feel important—and do it sincerely.

Dale Carnegie's "Six Ways" summarize what our engineering guests in this chapter have said. Combine the people wisdom you have just read from our experienced engineers with more depth on people skills by reading the books by Covey and Carnegie. Do both and you can be way ahead in life. But you'll be way ahead only if you make a habit of living the simple but elegant interpersonal skills that don't just come naturally to most people. Again, it's not what you know; it's what you do with what you know. The best among you will study these habits and become experts in genuinely making them part of your personality.

Chapter Summary: Takeoff Tips

- ❖ You will need to know how to communicate with people in order to be successful.
  - You may be on a design team, a plant floor, construction site, and so on.
  - You will also need to deal effectively with customers and others.
  - Your interpersonal skills will be as important as your technical skills.
- ❖ Live the Golden Rule: treat everyone like you want to be treated.
- ❖ Be friendly and respect everyone. Earn respect; give it and get it.
- ❖ Communicate assertively when there is a problem.
  - Don't hesitate to ask for clarification of your assignment.
  - Don't guess on an important matter.
  - Get to know the specialists who can answer your questions.
  - Save e-mails and important messages regarding project commitments.
- ❖ Focus on the project, do your best, but share credit with others.
- ❖ Learn people skills in the workplace on a co-op/internship assignment.
- ❖ As a manufacturing engineer, leadership is a key skill.
  - A genuinely enthusiastic leader gets more support.
  - Learn what you don't know, i.e., from machinists, welders, and technicians.
  - Teach what they don't know, i.e., solving engineering problems.
  - Use the talent on your manufacturing team. Share knowledge.
- ❖ Be prepared to do business with others outside your immediate function.
- ❖ Communicating is critically important: who, what, and how.
  - Who are you communicating with? Customer? Supplier? Boss?
  - What needs to be communicated? To you? By you?
  - How should you communicate? Use verbal and nonverbal techniques.
- ❖ Basics of communication:
  - Listen carefully, digest and analyze, then speak.
  - Have honesty, fairness, and high integrity at all times.
  - Again, treat everyone with respect; treat them like a peer.

- Make a point to have valuable two-way communication.
- Say what you mean and mean what you say. Commitments!
- Diversity: understand the types of people you are working with.

Exercises

1. Why are interpersonal skills important for you as an engineer?
2. This chapter features four engineers in different functions including testing, manufacturing, quality, and construction. What did you learn from each about dealing with people on the job?
   a. Danielle Buchbinder
   b. John Niblock
   c. Tanya Farrell
   d. Steve Perrigo
3. What are Dale Carnegie's "Six Ways to Make People Like You"?
4. Read Stephen Covey's *The Seven Habits of Highly Effective People* and Dale Carnegie's *How to Win Friends and Influence People.*

CHAPTER 23

Teamwork in Industry

WHY A CHAPTER ON TEAMWORK?

As an engineering student, you have been more or less involved in teams all your life—family, Little League sports, perhaps scouts, hopefully engineering clubs, and definitely laboratory partnerships. People working together . . . common sense, you might say. What's there to learn?

Team performance is very important on almost any job in industry, government, or academia. Your experience and instincts to date are probably not yet sufficient to meet a high-end employer's expectations as an excellent team member.

In fact, as an engineering student, you have spent a good deal of time in a non-team environment—in class, doing homework, taking tests, and so on. Also, the kinds of teams that you may have been on are most likely different than those that you'll be on in industry.

MOOG INC.—A TEAM-BASED COMPANY

As a sound business investment, Moog is a high-tech corporation that requires their employees, including engineers, to take teamwork training classes. According to their Web site, Moog designs, manufactures, and integrates precision control components and systems for military and commercial aircraft, satellites, space vehicles, automated industrial machinery, marine applications, and medical equipment. Moog's major business groups are: Aircraft, Space and Defense, Industrial, Components, and Medical Devices. To illustrate the dominance of teamwork at Moog, the Aircraft Group alone has a Site Leadership team, Integrated Product teams, Process teams, Product Service teams, Product Generation teams, Manufacturing teams, Assembly and Test teams, and numerous project teams. Moog invests in teamwork training for most all employees as a business necessity.

So let's learn more about teamwork in industry. And, who better to learn from than Peter Sergi, Barbara Davis, and Joost Vles, Employee Development managers from Moog, Inc. who teach teamwork to employees, including engineers, at the Moog campus. It's a privilege to have these experts with us to teach teamwork. What you are about to learn from these professionals will give you a head start in being an effective team member as you step into an industrial or other organizational environment, either as a co-op student or new employee.

INITIAL TEAM EXERCISE: BLIZZARD SURVIVAL

Mr. Sergi, Ms. Davis, and Mr. Vles distribute worksheets to each participant in a class describing a predicament where you and four friends are stuck inside your minivan in a blizzard in a remote area.* Your minivan's storage area includes a magnetic compass, a map of the immediate area, one pound of beef jerky, a few lengths of hose, a first aid kit, a sheath knife, a shotgun, a shovel, two blankets, a full one gallon gas can, and several hubcaps. There are seats in the van and in the glove compartment are a cigarette lighter, five pairs of sunglasses, and assorted maps. Each of you is wearing hiking boots and a wool jacket. The five of you have agreed to stay together. The time is 3 p.m. Mr. Vles leads the exercise:

- ❖ ***Step 1: Individual Ranking.*** On a grid chart, there is a listing of the 15 items listed above in the left hand column. Each individual in the class is asked, alone, to place the number "1" by the most important item for survival, "2" by the second most important, down through number 15, least important. Eight minutes is allowed for each person to do the ranking alone on his sheet in column 1 that is alongside the items.
- ❖ ***Step 2: Group Ranking.*** Then, the class is divided into teams. For the next ten minutes, team members discuss their logic for rankings and come to a group conclusion as to what the ranking order of importance should be for survival—items 1–15 in column 2.
- ❖ ***Step 3: Announcement of Survival Experts' Ranking.*** Mr. Vles reveals in column 3 what the consensus of outdoor survival experts' rankings are and why.
- ❖ ***Step 4: Bottom Line.*** The individual rankings are compared with the planning experts' rankings, then the team rankings are compared with the planning experts' rankings. In almost all cases the team rankings came closer to the experts' rankings. For brevity in this text, it is not important to explain the rankings and explanations. The key take-away is the concept of teamwork and its evidence of success.

So, what did we learn from this?

- ❖ Team performance generally beats individual performance, in this case by 80–85%.

---

*From Nicholas, Todd. Career Partners International, Buffalo Niagara.

- ❖ Team members need to actively participate by stating their views, attentively listening to the views of others on the team, and collaborating to come to the best conclusions.
- ❖ In a group, there may be someone with expertise of great value. Also, group discussion offers different perspectives, allowing consideration of more options.
- ❖ An exercise such as this, while it may seem far-fetched, develops real speaking, listening, and respectful collaborative skills that are necessary on an engineering project team. This exercise is not a waste of time; it's an eye-opener and a real skill builder for communication and teamwork.
- ❖ We should continue to pay close attention to what Mr. Sergi, Ms. Davis, and Mr. Vles will teach us about teamwork in this chapter.

COMMUNICATION—THE FOUNDATION OF TEAMWORK

Now Mr. Vles explains the importance of communication in teamwork. There is a communication loop* for us to share information as shown in Figure 23–1. It is simple to explain but not always easy to do:

- ❖ **Sending** seems easy. Anyone can send by written or spoken word. In person, nonverbal communication, including body language, is especially important, but also underestimated. Indeed, a UCLA study found that when it comes to communication, only 7% was absorbed through verbal content and 93% was taken away through nonverbal impact.
- ❖ **Receiving** also seems easy, but takes more energy than sending. It requires active listening and interpreting what is being received. Mothers tell us we have two ears and one mouth, and we should use them in that proportion. Think about it.

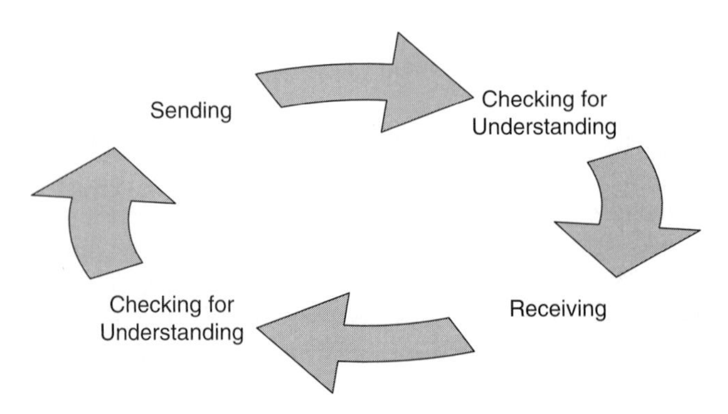

FIGURE 23–1 Communication Loop

*From Side by Side, Inc., 2010.

- **Checking for Understanding** is often the missing link in the communication loop. Too often neither the sender nor receiver checks for understanding. That's a sure recipe for misunderstanding, bad communication, and perhaps a project or relationship going off track. In good communication, especially on important issues, both sender and receiver have a responsibility to ensure that there is no misunderstanding.

In summary, effective communication is more than just sending, receiving, and checking for understanding. Effective communication means creating an environment where successful interaction will occur.

STAGES OF TEAM DEVELOPMENT

An effective team does not come together automatically. Teams evolve. Now, Ms. Davis explains this evolutionary process.*

- ***Forming.*** Forming is the initial stage of teamwork. Team members are getting to know one another. Not a lot gets done in way of tangible results, but it's necessary for the team members to become familiar and learn to respect each other. They learn their respective styles and values, as well as how they communicate. This is a relatively quiet stage.
- ***Storming.*** The honeymoon is over and we begin to see the first arguments. Now we know one another and we are anxious to get some things done. This is a hectic period—we don't know who's good at what or exactly how to communicate with one another. The conflict is frustrating, but there are better days ahead.
- ***Norming.*** This is a stage of agreement on how to work together and who's best at what functions. The team starts to click.
- ***Performing.*** The earlier stages of team development pay off. We now have a high performance team achieving the desired results. The performing stage does not last indefinitely. After all, teams are made up of people, meaning there's always the reality of conflict situations.

CONFLICT MANAGEMENT STYLES

Ms. Davis will now continue to tell us about conflict management styles. Individual personalities will tend to handle conflict in different ways. Here are some of the major conflict management styles.

- ***Accommodating.*** The accommodator says, "I don't like the conflict going on here, so I'm just going to give in. Do whatever you want, it's fine with me." The problem with this is that the conflict may disappear

---

*From Tuckman, Bruce Wayne. "Development Sequence in Small Groups," 1965.

for the moment, but there may be unresolved ideas that were removed from the table and pushed under the rug.

- ❖ *Competing.* Competitor #1 says, "I want my idea first." Competitor #2 says, "No, I want my idea first." Competitors are only worried about their own stuff. At the end, someone will win, but is it a win-win? Probably not.
- ❖ *Avoiding.* As with the accommodator, avoiders don't like conflict. Their avoidance strategy is to just step aside and let the team proceed without them. The team misses out on a potentially valuable contribution from that person and the avoider misses the opportunity to be part of the solution.
- ❖ *Compromising.* In compromising, we give a little and take a little. Conflict is reduced or avoided, but we've all given something up, and that is the opportunity to achieve the best solution. Compromising is better than the previous approaches in that all of the ideas have been considered, but there's an even better teamwork strategy.
- ❖ *Collaborating.* When we collaborate, we shoot for the win-win solution. We try to all come together positively, as partners, for best results. We discuss the different viewpoints openly and respectfully, building on each others' honest inputs and bringing the best ideas into one solution package.

THE PROCESS OF BUILDING COLLABORATION

We have established that collaborating toward win-win solutions is the best teamwork approach. Now, Mr. Sergi will explain the process flow of building collaboration,* i.e., how we get there.

1. *Manage Your Feelings.* Anger is a natural human emotion, particularly when your personality or idea is contrary to other team members. Potential anger needs to be understood, along with the fact that exploding in anger can be very destructive to future positive progress as a team. If you or another team member is inclined to have a fit of anger, the best approach is to cool off, perhaps walk away. Prevent potentially irreparable damage to team relationships by working out differences as soon as a calm and rational approach to the problem can be used.
2. *Create a Supportive Climate*
   a. *Coming Together on Mutual Ground.* Especially when there is expected conflict, it's important to meet in a physically or psychologically neutral space. Allow time to get comfortable with each other—don't feel rushed.
   b. *Having a Sense of Openness.* Respect others on the team. Listen to and value others' ideas and concerns. A great step toward collaboration is if each team member has an attitude of honest, anger-free openness. Openness is also a great step toward eliminating bottled-up stress that can be harmful to your health and relationships.

*From Side by Side, Inc., 2010.

3. ***Focus on the Facts.*** If there is conflict, try to make emotion irrelevant. Don't accuse. Don't make assumptions. Don't jump to conclusions. Instead, lay out the facts and examine them objectively. Don't describe the emotion, describe the facts. This will lead to a more constructive, friendly conclusion.
4. ***Describe the Goals.*** At this point, we need to check for understanding. Where are we going? Everyone on the team needs to be pulling together toward the same goals. Many unskilled teams don't check to see that they're all heading for the same target. A team leader will check for understanding by summarizing the goals. One or more team members may paraphrase the summary just to be sure there is understanding of the desired outcome. If there's agreement, then we perform the final step.
5. ***Create Solutions.*** Now we have a collaborative team, on the same wavelength for problem solving with agreed-upon objectives. Everyone knows the goals and now we can brainstorm the best solutions. Since the team now has trust and can work together, wild, creative ideas can be suggested—as many ideas as possible. No one gets shot down for making a goofy suggestion. Everyone contributes on an uninhibited basis. Then we work together to narrow down the ideas and decide which are best suited to reach our goals. Teams operating in positive, creative environments such as this have proven to come up with the best results. Creating solutions is what win-win teamwork is all about, and better solutions are why organizations value teamwork so highly.

INDIVIDUAL ESSENTIALS OF TEAMWORK

So far, this chapter has described teamwork collectively. But, of course, teams are made up of individuals. Now, Mr. Sergi, Moog employee development manager, will conclude our teamwork chapter by describing individual qualities necessary to be a good team member. Regardless of your technical skills, these elements are crucial for your success on a team or as an individual contributor. Here they are—have these qualities and be successful.*

- ***Energy.*** People like those who have energy. Accomplishment and success require energy. The person who mopes around is a drag: personally and to the organization. You have a behavioral choice on this, as with the items to follow.
- ***Enthusiasm.*** Ralph Waldo Emerson said, "Nothing great is accomplished without enthusiasm." Your enthusiasm is contagious and can allow you to be a team leader. If you have a great idea and are not enthusiastic about it, no one's going to buy your idea.
- ***Passion.*** Successful people choose products and functions that they are passionate about. They love their work, including the customers they

*From James Tressel, Head Football Coach, Ohio State University.

serve. If you don't have passion, it shows, and in that case you may belong somewhere else.

❖ ***Punctuality.*** You have to be on time. You have to show people that you're reliable, and will be there. Whether it's for a team meeting or any other commitment, there is no excuse for being late.

❖ ***Integrity.*** Defines your character. Honesty and integrity are fundamentals for your good reputation, both professionally and personally.

❖ ***Good Behavior.*** Be a good worker and a good team member. Be polite and try to get along with people. Get in there and do the best you can.

❖ ***Understanding.*** Show an interest in others and listen to them. This will allow you to know what they think and why they act as they do. Positive understanding of people and their motivations will strengthen your relationships and abilities as a leader.

❖ ***Respect.*** Everyone wants it. How do you get it? Show respect. If you don't show respect, it'll ruin relationships and perhaps your career. Show respect to everyone you work with at all levels. That will go a long way toward your success.

Ms. Davis, Mr. Vles, and Mr. Sergi have just explained the fundamentals of teamwork and being a good team member. Despite the apparent common sense nature of this subject, teamwork requires real skill. Put the teamwork instruction of these employee development managers from Moog into action along with your technical skills, and you'll be on your way to top performance on the job. You can do it!

Chapter Summary: Takeoff Tips

❖ Team performance is very important on almost any engineering professional job.

❖ Lessons learned from the Blizzard Exercise:
- Team performance generally beats individual performance by 80–85%.
- Team members need to actively participate both listening and speaking.
- The team may include someone with expertise of great value.
- Group discussion offers different perspectives with more options.

❖ Communication is the foundation of teamwork. The communication loop:
- Sending: by written or spoken word. Also nonverbal body language
- Receiving: seems easy but requires active listening and interpreting
- Checking for understanding: both sender and receiver are responsible

❖ Stages of team development:
- Forming
- Storming
- Norming
- Performing

- Conflict management styles:
  - Accommodating
  - Competing
  - Avoiding
  - Compromising
  - Collaborating
- The process of building collaboration:
  - Manage your feelings
  - Create a supportive climate
  - Focus on the facts
  - Describe the goals
  - Create solutions
- Individual essentials of teamwork:
  - Energy
  - Enthusiasm
  - Passion
  - Punctuality
  - Integrity
  - Good behavior
  - Understanding
  - Respect

Exercises

1. Why do we include a chapter on teamwork?
2. Describe the communication loop.
3. Explain the stages of team development.
4. Describe the conflict management styles.
5. How does a team create a supportive climate?
6. What are the individual essentials of teamwork?

CHAPTER 24

Leadership

Leadership, and its close cousin, management, involve getting results accomplished through other people, as opposed to being an individual contributor.

More than one student has come to me, after becoming president of an engineering student club, with a problem. That problem was that they were doing all the work. Other members of the club were not involved. The result? Those student presidents were stressed by balancing the work of their club along with their engineering coursework. Their grades were starting to suffer, as was the enthusiasm of their club membership. These club presidents were trying to do everything themselves, which is an impossible task, while overlooking the potential talent and commitment of other members of their organizations. What was needed? The answer is that these presidents needed to become leaders and managers rather than individual contributors. Basically, here is the solution that these new presidents carried out and, to the benefit of all, worked every time.

PLAN. ORGANIZE. CONTROL. LEAD.

These are the fundamentals of leadership and management. Have a plan, delegate authority and responsibility, and have control steps to see that the right things are being accomplished on schedule, and lead through communication and motivation. Sounds simple. The foundation principle of leadership and management is getting your organization involved in reaching a goal. This means getting people and resources committed to attain a desired result. Leadership includes managing the organization so that everyone is working together. Members are involved. They have responsibility. They are motivated. They feel important and committed to reaching goals. Together, there's much higher morale and enthusiasm to make things happen. In short, there's good leadership in action that involves everyone.

FUNDAMENTAL LEADERSHIP/MANAGEMENT STEPS

Planning

This means having a vision, a mission, and strategic objectives for the organization. It means writing down action steps to reach each objective. As indicated in Chapter 20, written action plans include an objective, specific results expected, and a series of action steps noting the person accountable and completion date alongside each step. In short, planning specifies what needs to be done, when, and by whom. Budget and quality standards are also an important part of planning.

Organizing

This means setting up a structure so that the plan can be accomplished. Keys to effective organizing are:

- Knowing your objective.
- Deciding what people and other resources are needed to accomplish your objective.
- Delegating responsibility and authority to those who can do the job. When delegating, it's important to define the job objectives, duties, responsibilities, and authority that goes with the job.
- Ensuring that the person to whom you're delegating responsibility is willing and able to do the job. This includes participative management, i.e., two-way communication with the person receiving delegation, such that there's clear understanding of what's to be accomplished and the exchange of any thoughts about goals and restrictions. Participative management is practiced by effective leaders to include involvement with all members of the organization. Everyone is valued and encouraged to contribute their ideas for improvement of the operation.
- Ensuring coordination between organizational units.

Controlling

This management step exists to be sure that the action plan and organization are making the intended progress toward the goals. In industry, you would

- *Set Performance Standards.* The team needs to know what's going to be measured and how. They need to know the level of quality, cost, time, and any measurable element that will contribute to the plan's success.
- *Establish Control Checkpoints.* This ensures the team is on schedule and within budget. Define boundaries, including acceptable limits and warning controls, to let the leader or team know that something is wrong.
- *Measure Performance Against These Standards*
- *Take Steps to Improve Performance When It Deviates from the Standards.* This includes taking action in time to correct the problem before it becomes worse.

Leading

This broad management function includes planning, organizing, and controlling that have just been summarized. Moreover, leadership rises above specific management duties. As we'll see later in this chapter, great leaders are able to inspire people to come together and achieve results that they would never have believed possible. In the planning stage, great leaders are able to instill a vision and mission that those being led will embrace with enthusiasm and commitment. Specifically, effective leaders, beyond the management functions just described, are especially skilled in communicating and motivating.

Communicating

Leaders communicate the big picture. They also get into specifics. This means transmitting and receiving required information. Effective communication ensures that everyone understands the plan and the organized steps designed to achieve the desired result. In effective communication, it is important that:

- Leaders clarify and confirm details of the plan and organization to everyone involved. It's important that leaders encourage two-way communication so that there's understanding and acceptance of the plan, organization, and controls.
- Team members receive communication from leaders and each other with ideas to optimize results. Team members should ask any questions required for clarification of what is to be done and make suggestions for optimum results.
- Everyone—up, down, or across the organization—transmits and receives ideas as necessary to accomplish the goals. Transmitting is done through the written and spoken word. Receiving is done by listening and reading. Everyone's ideas are important and need to be valued. It's a leader's responsibility to be sure that everyone in the organization is aware of this value and acts on it. A culture of excellent communication is part of every outstanding organization. The most successful leaders embrace the humility principle, "I don't know everything there is to know about anything." They surround themselves with people who see things differently and encourage different perspectives to be shared openly. The great leader enables the translation of different viewpoints into one unified perspective that provides the best possible solution or direction for a team or organization.

Motivating

Leaders motivate people. They do that through their personal enthusiasm about the purpose and goals of the organization. Moreover, leaders create an environment of empowerment. What's empowerment? Empowerment means that members of the organization feel important and responsible for their part in carrying out the vision, mission, and strategy communicated by the leader. In groups that feel empowered, each individual is likely to be turned on and enthusiastically committed to the organization's success. Team members are committed to do their own job well and to make any suggestions that will benefit the organization as a whole. Great leaders excite team members by describing an unsatisfied need, suggesting a course of action, and encouraging the team toward a more desirable goal. The role of a leader is to create incentives and enthusiasm so that the team will be motivated and personally empowered. This motivation will get everyone moving toward goals of the organization. Motivation of individual efforts toward the common goal of the team is the ideal. The closer a leader comes to achieving this desired state, the better the leader.

LEADERSHIP IN INDUSTRY

Now, we will benefit from the wisdom of two experts from industry, William Danesi and Robert Strassburg, who have considerable experience in engineering leadership. Bill Danesi received his BS degree in Chemical Engineering from Texas A&M University. He has 35 years of engineering management experience at Xerox, General Electric, and Scott Aviation corporations. Bob Strassburg has BS and MS degrees in Civil Engineering and has been an engineering associate and project manager at Praxair Corporation. He is now CEO of Trinity Leadership Consulting, LLC, a leadership consulting firm. Together they'll address the subject of engineering leadership based on their expertise, with examples that you can use.

Bob Strassburg opens with an invitation for you to understand engineering leadership as well as leadership in more general terms. Bill Danesi adds that every year, there's an increasing need for young engineers to step up and assume leadership roles in enterprise—industrial, academic, or government. There'll be many leaders/managers retiring soon, so a next generation of leaders is required. Adding to this fact, there's an age gap in many organizations between the leaders and those of you who are just starting your careers. Therefore, as a graduating engineer, you're in a great position to assume necessary engineering leadership responsibilities in the United States and throughout the world.

Leadership Defined

Engineering leaders will need to prepare their people for a world of rapid change and opportunity. So, Mr. Strassburg will define leadership, first through the Merriam-Webster dictionary.

- To **lead** (verb)—means that you are taking on the responsibility to get something done, through somebody else, and taking action to produce a desired result.
- A **leader** (noun)—is a person who has commanding authority or influence. It means that you can exert some sort of influence over someone else to produce results.
- **Leadership** (noun)—is the office or position of a leader. It is also the quality of a leader and the capacity to lead.

Looking at the above definitions, we need to recognize the difference between the nouns and the verb. Often, a person is given the title of leader but fails to take action and actually lead. The nouns leadership and leader define the desired state. The verb to lead defines action. The key is to recognize that the leaders need to lead in order for leadership to exist. That may seem obvious, but it's a fundamental that is too often overlooked. Leaders need to get desired results through other people. Action and results are what matter.

Leadership is first and foremost about people. It's about motivating people to get something done. It's about taking on the accountability for something that someone else is going to complete. Leadership is about success. These days especially, leadership is about change; it's about the ability to lift people up and work through change, in spite of the fact that we're living in a dynamic and uncertain world.

Leadership Example: President John F. Kennedy

Mr. Danesi invites us to look at President John F. Kennedy as an example of a person who understood the power of his position, and the ability to communicate an audacious vision to the nation: going to the moon. In 1961, President Kennedy said, "We choose to go to the moon. We choose to go to the moon in this decade and do the other things, not because they are easy, but because they are hard, because that goal will serve to organize and measure the best of our energies and skills, because that challenge is one that we are willing to accept, one we are unwilling to postpone, and one which we intend to win."

The key in this example is that President Kennedy, through this powerful speech and follow-up commitment, was able to achieve the goal by 1969. Despite his assassination in 1963, the dream and action plan lived on. We not only reached the moon within the scheduled time commitment, other huge advances were made in computers, aviation, astrophysics, telecommunications, robotics, medicine, bio-engineering, astronomy, and the list could go on. While reaching the moon was President Kennedy's specific objective, he no doubt realized that many other technical breakthroughs would also be achieved through the combined efforts of industry, the government, and universities. No one in 1961, including President Kennedy, knew whether it was actually possible for Neil Armstrong, Buzz Aldrin, and Michael Collins to fly to the moon by 1970, but Kennedy's inspirational leadership challenge was the starting point of making it happen.

President Kennedy was not the only one to do something like this. During the 20th century, John Kennedy was the second of two popular, eloquent, farsighted U.S. presidents who proposed such a massive successful undertaking that involved new frontier technology. The first was President Franklin Delano Roosevelt, who launched the secret Manhattan Project after Albert Einstein explained to him that, theoretically, enormous energy could be obtained by splitting the atom. Both Germany and the United States were racing toward such technology to win World War II. The United States got there first with the atomic bomb.

During the 21st century, will another farsighted U.S. president announce a challenge to make Wilson Greatbatch's prediction of nuclear fusion, described in Chapter 17, pages 173–174, a reality to solve our energy crisis? How soon might such an inspirational energy challenge announcement happen? Will we ever run out of oil? If so, how soon? How many who are reading this page will take technical or political steps toward making nuclear fusion, or other energy solutions, a reality? Is there any doubt that you as an engineer are critically important to solutions and opportunities for the future of all? No, there is no doubt!

Now, let's get back to everyday leadership issues. Mr. Strassburg will tell us about a common problem: a manager who is not a leader.

Management Without Leadership

It's not uncommon in a technical organization that a person who's a great engineer gets promoted to management, despite the fact of not being qualified to be a leader. The presumption was that the person, being technically excellent, would be an equally excellent leader. Not necessarily so: we might have just lost our best technical contributor and gained an incompetent manager. Mr. Strassburg tells us what happens when a non-leader becomes a manager.

- ❖ ***Non-Leaders Are Reactive.*** That is, they carry only their specific technical focus into their management role. With narrow vision, they react to events, rather than leading and anticipating situations. This causes poor response in crisis situations.
- ❖ ***Non-Leaders Wonder What Happened.*** Leaders make things happen; non-leaders wonder what happened. The non-leader who has not been trained in management has not done management planning, leading, or monitoring to see what's happening with their team. No wonder they're surprised when a crisis occurs.
- ❖ ***Non-Leaders Want To Maintain Control and the Status Quo.*** Insecurity causes them to want to hold all the cards and play them themselves. They're afraid of change and are risk averse because they fear sticking their neck out and being responsible if something goes wrong. They withhold information, both inside and outside their teams, as a means of controlling and manipulating circumstances to their benefit. Non-leaders are reluctant to encourage and develop competence and accomplishment within their teams that are greater than their own.

In short, this kind of person is ineffective as a manager and does not qualify as a competent leader.

MANAGER AND LEADER: AN IMPORTANT DISTINCTION

We often think of a manager and a leader as one and the same. Managing and leading are both important functions, but there are some distinctions worth noting. Mr. Strassburg describes the characteristics of an engineering leader versus an engineering manager.

The Engineering Leader

- *Is Proactive.* The leader is forward thinking and takes responsibility. Being proactive, the leader chooses appropriate action, rather than reacting to events.
- *Has Vision.* The leader is able to imagine something in the future. Moreover, leaders are able to communicate that vision to the people in their organization.
- *Creates Alignment.* That is, leaders get everyone pointed in the same direction. People being led understand the vision and purpose of the organization and everyone is moving effectively toward the same goals.
- *Promotes Growth.* Leaders want their people to develop. They want their people to accelerate their learning. They want them to become more competent, even more competent than they are.
- *Promotes Continuous Improvement.* This goes along with the idea of growth. Both the organization and individuals within it need to get better and better, for the benefit of all.
- *Prevents or Puts Out Fires.* They also assist other people in doing that.
- *Leads by Example.* They model the behavior that they want their people to exhibit.

The Engineering Manager

The engineering manager may or may not be a leader. Ideally, the manager is also a leader. By contrast to the leader, the engineering manager should have these minimum qualifications:

- *Is Responsive.* The manager's always looking out for fires and is ready to respond when necessary—as opposed to leadership fire prevention.
- *Stays Focused and Maintains Control.* They monitor what's going on through measurement and status reports and are ready to respond when things get outside of acceptable control limits.
- *Provides resources* as necessary.
- *Manages the firehouse* using traditional and/or unconventional management practices.

THE LEADERSHIP MODEL

Mr. Danesi and Mr. Strassburg collaborate to explain both the intrinsic characteristics of a leader and the developmental characteristics of a leader.

- *Intrinsic Characteristics.* They include the personal traits of the leader. Intrinsic characteristics are embedded in the personality and accompany the person who's a natural leader who has recognizable confidence, people skills, and other leadership talents. These are the personal traits and qualities of a natural leader:
  - Visionary, sees and grasps opportunities
  - Natural skill in project management

- High respect for people, willing to delegate
- Trust, ability to build trust and build powerful relationships
- Flexible and oriented to change
- High initiative and enthusiasm
- Charisma and persuasiveness
- Organization and discipline
- High personal and professional integrity
- Infectious confidence toward a successful outcome

❖ *Developmental Characteristics.* They are those specific skills of a leader that can be developed through training.
- Communication skills
- Conflict resolution skills
- Technical skills
- Planning skills
- Organizational skills
- Entrepreneurial skills
- Administrative skills

Whether you're a natural or can grow to be an effective leader through training, Mr. Danesi and Mr. Strassburg encourage you to take every opportunity to learn, practice, and use these developmental skills. They'll serve you well whether you're an individual contributor or a future manager.

CONFLICT RESOLUTION

One major leadership skill that can be developed is conflict management resolution. Conflict has been addressed in Chapter 23, but in view of its importance it's valuable to have Mr. Danesi and Mr. Strassburg's instruction in this chapter also—especially since this is conflict resolution from a leadership perspective.

Conflict is defined as "the behavior of an individual or a group that impedes another party from attaining its desired goals." Anticipation and resolution of conflict is an important leadership responsibility. Conflict arises when there's a collision between people or groups with different perspectives. Different perspectives can be good in a creative process, such as brainstorming. But when negative conflicts aren't resolved, things can go bad, and here's where leadership comes into play. The best leaders anticipate harmful conflicts.

Types of Conflicts

Conflicts can involve manpower, resources, equipment, facilities, capital spending, costs, or administrative procedures. Some of the biggest areas of conflict in a technical organization include schedules, technical opinions, and personality clashes. Any of these potential sources of conflict can stand between people as they try to accomplish something together.

As a leader, the best way to avoid conflict is to prevent it from happening to begin with. Personality clashes can be reduced by managers knowing their people and who not to mix together. Also, you need to manage relationships to see that conflict doesn't get out of hand.

Methods of Conflict Resolution

There are various behaviors a leader can use to resolve conflicts.

- *Withdrawing.* Getting everyone to step back for some period of time to cool off, especially when emotions are running high.
- *Smoothing.* A leader smoothes by asking the group to calm down and telling them that we need to think this through and consider the other person's point of view.
- *Compromising.* Here, the leader tries to find balance between the different points of view so that everyone gets something of what they want.
- *Forcing.* Sometimes there's an impasse and a negative business consequence for the conflict to continue. In such a case, a leader may need to step in and say, "This is the decision; this is the way we're going to go."
- *Confronting.* On other occasions, a leader may need to get in people's faces and say, "What you're doing is inappropriate; we need to get this resolved and we're going to sit down and figure this thing out."

Managing conflict is one of the most important responsibilities of a leader. Good leaders listen for signs of conflict and deal with the situation promptly. Conflict resolution calls for leaders to use their effective communication skills—listening, speaking, and especially having a keen understanding of people and their motivations.

THE IMPACT OF LEADERS ON THE ORGANIZATION

In addition to the organization being successful, good leadership has other advantages, all of which contribute to the success.

- *People Feel Significant.* Good leaders respect their people and make them feel significant. When a person feels significant, when they feel that they're an important part of the organization, their self-esteem increases. They're empowered and motivated. They're committed to the organization's goals. If your boss thinks you're outstanding, you won't want to disappoint that leader or, especially, yourself.
- *Competence and Learning Increases.* Leaders with enthusiasm and respect for everyone end up with more competent people who are anxious to improve themselves and the organization. Strong leadership inspires people to do their best and continuously learn. Creativity increases throughout the organization.
- *A Community Exists.* The organization has high morale. They feel like family. That makes conflict problems much less likely. Teams are effective. Turnover is low and loyalty is high.

❖ *Work Is Exciting.* You have every reason to believe that your engineering career will be exciting. You can choose work that you're really good at and enjoy. You'll value the importance of your product or process. And, when your leader is inspiring and thinks you're great, you'll come to work with a smile and a spring in your step. Leadership rubs off—you may already be a leader!

Chapter Summary: Takeoff Tips

❖ Leadership and management involve getting results through other people.
❖ To become leaders and managers, rather than individual contributors:
  - Plan: have a vision and written objectives for the organization
  - Organize: delegate responsibility and authority
  - Control: ensure the plan and organization are on course
  - Lead: perform the above functions, plus communicate and motivate
❖ Leadership is first and foremost about people. Specifically, it is:
  - Motivating people to get something done
  - Being accountable for what someone else is going to complete
  - The ability to lift people up to work through uncertain change
  - Success, through other people
❖ Management without leadership:
  - Non-leaders are reactive with narrow vision, they react vs. leading
  - They wonder, what happened?
  - They want to maintain control and the status quo
❖ The engineering leader:
  - is proactive, forward thinking and takes responsibility
  - has vision and imagines something in the future and communicates it
  - creates alignment by getting everyone pointed in the same direction
  - promotes growth and wants their people to develop competence
  - promotes continuous improvement of the organization and individuals
  - leads by example, modeling desired behavior
❖ The leadership model:
  - intrinsic characteristics: the personal traits of a natural leader
  - developmental characteristics: can be taught through training
❖ Methods of conflict resolution:
  - Withdrawing
  - Smoothing
  - Compromising
  - Forcing
  - Confronting

❖ The impact of leaders on the organization:
  - People feel significant
  - Competence and learning increases
  - A community exists
  - Work is exciting

Exercises

1. What's the objective of leadership and its cousin management?
2. Explain the four common fundamentals of management and leadership.
3. Why is leadership first and foremost about people?
4. Explain why President Kennedy is used as an example of leadership.
5. Explain the difference between a manager and a leader.
6. What are the intrinsic characteristics (natural traits) of a leader?
7. Name the leadership qualities that can be developed through training.

CHAPTER 25

Effective Writing and Presentation Skills

INTRODUCTION

Beyond your engineering technical proficiency, a key element of your professional success is your communication ability. This chapter focuses on effective writing and presentation skills. There's a lot of information in this chapter, but you'll do well to pursue additional study on this subject.

Much of the information in this chapter is applicable to writing resumes and cover letters (Chapter 5), interviewing and follow-up (Chapters 8 and 9), and leadership (Chapter 24). More broadly, however, it's important to emphasize, up front, how crucially important communication is to you and your career. Effective communication, including writing, presentations, and interpersonal skills will have a profound impact, for better or for worse, on your career.

You're finishing your co-op or internship assignment and need to write a project report for your faculty advisor; your grade depends on a well-written report in addition to what you did technically. You have a bright idea for a technical concept that you would really like to have your company invest in for development, but your idea won't be funded unless you can effectively communicate the benefits of it to management. We can name countless other reasons that a good engineer or engineering student needs to have effective writing and presentation skills. That's what this chapter is about.

Many engineering students might place report writing and giving an oral presentation rather low on their list of things they must be able to do. But . . . Fact: The abilities to write well and give a good presentation are critically important skills for a successful engineering career. You have heard it consistently from our engineering experts in previous chapters; those who communicate well are much more likely to be hired, promoted, and able to be a total engineer in any function, including R&D, design, project management, sales, production, and line management. So, three realities to consider:

1. You need effective writing and presentation skills.
2. Once you have these skills, you'll enjoy using them and appreciate their value.
3. Your communication skills will lead to a more rewarding career.

For several years, William Grunert has been conducting technical communication seminars and workshops for University at Buffalo engineering students as well as for engineering companies. He has BS and MS degrees in Mechanical Engineering from the University of Notre Dame, completed a 35-year career in engineering management at Praxair Inc., and has been a lecturer in technical communications for 12 years.

During his career at Praxair, Mr. Grunert saw countless examples where the effectiveness of communications directly correlated to how successful an endeavor was or, conversely, how problems could develop in the execution of the work. Today, when asked to comment on the importance of communication in the workplace, he offers the following observations:

1. With a computer on every desk, engineers and technical professionals are required to communicate directly and frequently with both internal and external customers. The quality, accuracy, and clarity of their reports, presentations, and general documentation directly impact the success of the project.
2. An improved technical communication skill is the necessary companion to increased technical knowledge and depth of understanding. Technical professionals with strong communication skills present and sell their ideas more effectively; thus, they enhance their organizations' effectiveness and are perceived as being truly professional.
3. Employers and professional engineering organizations continually cite poor communication skills as a major competency gap that manifests in a variety of costly ways: unproductive teams, customer-supplier misunderstandings, and customer dissatisfaction. A 2004 survey of 120 American companies, conducted by the National Commission on Writing, concluded that one-third of the employees in these companies write poorly. The problem shows up not only in e-mails but in other reports and documents as well.
4. In my workshops with engineering companies, I ask them to complete a questionnaire about their workplace writing. On average, these workshop attendees spend 25–30% of their time writing. When you factor in other workplace communications (meetings, occasional presentations, and one-on-one communications), engineering professionals may spend well over half of their time communicating. The importance of doing it well is compelling.

Mr. Grunert begins a communication class with the observation, "When you are on an engineering co-op assignment, you will have to write a report and perhaps give an oral presentation at the end of your work period, describing your assignment and what you accomplished. How many of you engineering students view that as an opportunity?" Very few hands are raised. He continues, "How many view it as something you have got to do?" Most hands go up. Mr. Grunert then suggests that you adjust your thinking to view engineering courses that emphasize communication as an opportunity. He goes further to suggest that one of your objectives in your engineering education should be to develop your skill as a very effective and professional communicator.

OVERVIEW OF EFFECTIVE COMMUNICATION

Communication Objectives: Decide whether you're writing:

- ❖ ***To inform?*** You have information and you need to get it out to people. Technical communication is all about conveying information.
- ❖ ***To evaluate?*** You've done an evaluation. Why is it important? Describe it. What are the results of the evaluation?
- ❖ ***To persuade?*** You're writing a proposal to your management about your new idea. What do you want to have happen? Now, you're writing to persuade your boss and others in the management team to fund your new idea. Your effective, logical, persuasive writing will determine whether your new idea ever gets off the ground.

Your objective determines your content. So, the writer's challenge is to:

- ❖ Recognize the importance of effective communication.
- ❖ Define expectations for your readers and for you.
- ❖ Produce high-quality content in your report. It's key for a good document.
- ❖ Package information in easily digested pieces.
- ❖ Produce documents that are logical, accurate, clear, concise, and easy to read and understand.

Good writing is coherent, i.e., clear, logical, and understandable. Good writing lacks confusing "noise." Noise makes written material difficult to read and understand and occurs when:

- ❖ A document is poorly organized.
- ❖ Ideas are sequenced illogically.
- ❖ Sentence and paragraph structure are poor.
- ❖ Format is lacking.
- ❖ Language and grammar are poor.

Let's illustrate by looking at two examples of a memo: one is poorly written and filled with noise and the other is written in a much clearer and more understandable way.

Please read the memo in Figure 25–1, and answer the following questions.

- ❖ What is the purpose? Is it clearly stated?
- ❖ What is the message? Is it clear? How is it delivered?
- ❖ How would you revise this memo?

MEMORANDUM

To: Whom it may concern
From: J. Doe, Plant Utilities Engineer
Subject: ABC Corp. compressor replacement parts

General Hardware, Inc. has been our sole source supplier for the last 15 years because of the following reasons: ABC Corp. has chosen to destroy all prints for compressors manufactured prior to 1950. Our ABC compressors were manufactured in 1931. ABC had also priced themselves out of the market by charging five times what we currently pay for compressor parts for sparse remaining inventory. We have tried other companies, but their quality was so poor that we were not able to use the parts. On many occasions we have to supply a sample so a mold could be made for the item to be formed in. This process from other suppliers inflates the costs and makes them uncompetitive. I know of no other company that can meet our demand for high-quality compressor parts in a timely manner. We are always open to consideration of other vendors, should they meet our requirements. I respectfully request that the order for compressor parts, at a cost of $3737.98, be processed without delay. Thank you for your prompt attention to this matter.

FIGURE 25–1 Ordering Spare Parts Memo Version 1

Now, could there be a memo with better organization and focus, better sequence of information, and improved style and readability? Please review the revised memo in Figure 25–2, and compare with Version 1 in terms of purpose, message, and understandability.

Let's summarize conclusions from the two sample memos. In any written material, a poorly organized document presents a problem for the reader to understand and take action. The first memo is an example of poor organization, including wrong sequence of information. Unnecessary detailed information was introduced ahead of the main point. There was no paragraph structure. There was no format. Let's now examine these deficiencies in more detail, and discuss how you can overcome them in your writing.

- ❖ ***Inadequate Organization and Focus.*** In Version 1, the writer has not taken the time to organize the information and focus it in a way that's readily understood by the reader. A good report is physically well structured and focused. It is focused to the reader of the report with the reader's interests and needs in mind. The reader can scan the report to readily grasp the big picture, as well as the important details. Paragraphs are well structured. Bullets are used to highlight a listing of necessary information. Unnecessary information is eliminated. The writing moves forward, actively toward a logical conclusion crystallizing what happened or what needs to be done.

MEMORANDUM

To: Purchasing Agent
From: J. Doe, Plant Utilities Engineer
Subject: Purchase of ABC Corp. Compressor Replacement Parts

Please immediately purchase the subject ABC Corp. compressor replacement parts (see attached specification) from General Hardware Inc., at a total cost of $3737.98. We have again selected General Hardware, our sole supplier of these parts for the past 15 years, for the following reasons:

- To my knowledge, General Hardware is the only supplier who can provide acceptable ABC compressor replacement parts at reasonable cost.
- ABC no longer services their older, pre-1950 compressors (our units were made in 1931), and its replacement parts prices are five times higher than General Hardware prices.
- In the past, we have had no success with other suppliers' parts due to high cost and unacceptable quality.

If possible, we should develop a qualified alternate supplier for these critical replacement parts. This would give us greater assurance of obtaining these parts in the future. We welcome your suggestions.

I will call you on May 1 to confirm your placement of the order and to schedule a meeting to discuss potential alternate suppliers. In the meantime, please call me if you have any questions.

FIGURE 25–2 Ordering Spare Parts Memo Version 2

❖ ***Bad Sequencing of Information.*** In Version 1, information is sequenced badly, scattered, and disorganized with important information buried someplace where the reader might miss it. It's a chore to read and understand. A good report has a logical sequence of information. The most important information should be stated up front where it can't be missed, especially if action is required. Appropriate details can be included in the middle of the report or memorandum, perhaps with bullets highlighting information as indicated above. The end of the report gives a conclusion and perhaps a call to action.

❖ ***Ineffective Style, Language, and Readability.*** In Version 1, the writer may have thrown some material together without thought of organization, focus, or sequencing of information. Apparently, no thought was given to format or paragraph structure. Moreover, the writer who has ineffective style and language has probably not focused on the reader and the reader's ability to understand the report or interest in taking action on it. A good report is organized, focused, and sequenced so that the reader will be interested and will understand it. A good report is written so that the reader will immediately know the situation, details, or action steps, and will grasp an appropriate conclusion. The style and language of a good report are professional with good grammar and always with the reader of the report in mind. Motto for a good writer—"Be reader-focused."

In summary, good writing requires that you:

- Have a thorough knowledge of the subject
- Focus on the readers
- Know what to communicate
- Think clearly and organize logically
- Use clear, understandable language

THE THREE P's OF TECHNICAL WRITING: PLANNING, PROCESS, PACKAGING

These are key elements of good writing that will enable you to produce a document that is easy for you to create and easy for the reader to understand.

Planning

Think and plan before you write. Ask yourself these five key questions:

1. Why are you writing?
2. Who are your readers? Target your audience. Understand them. In a co-op assignment your audience could be your employment supervisor as well as your faculty advisor. Throughout your engineering career, your readers could be almost anyone: management; team members; customers; and either technical people, non-technical people, or a mixture of both. Emphatically, be reader-focused. If you're communicating with top management, they'll be especially impressed and appreciative when they receive a well structured, easy to read report that captures their interest in your proposal.
3. What are your readers' expectations? What response do you want? In a co-op assignment, one of your desired responses is probably to get a good grade. That means your readers will expect you to give them a well-organized, readable, report of what you did, the quantitative results you achieved, and how the results will benefit the employer.
4. What information should you provide? What level of detail? A big part of planning before you write is considering the above questions of why you're writing, who your readers are, and what's the response you want. Answering these questions will guide you as to how much information you should include in your communication and what level of detail is required.
5. How do you deliver your message? Report? Memo? Letter? E-mail? Recognize that any of these options could be appropriate, depending on your purpose. You might be sending a quick, but well-organized e-mail or memo, or it could be a full report of a co-op assignment or a proposal of your new idea that you want to get funded by management or a customer.

Process Steps in Communication

Communication is, indeed, a process. It's very much like an engineering project for designing a machine or a plant. Here's an outline of the steps. We'll go into more detail later—but it's good for you to have the big picture now.

- ❖ ***Plan.*** Answer the five key questions above.
  - Define your purpose, understand your audience, and focus your report.
  - Define your content. Select information based on audience needs and expectations.
  - Define appropriate level of detail.
- ❖ ***Organize.*** Use an appropriate organization model. Organize your information logically, using an outlining technique to help you.
- ❖ ***Write.*** Develop a content draft, using clear understandable language.
- ❖ ***Format.*** Integrate text and graphics, formatting for easy grasp and understanding.
- ❖ ***Edit.*** Review for language, readability, and grammar.
- ❖ ***Present.*** Plan and organize your presentation.
  - Prepare effective visual aids.
  - Deliver your presentation.

We strongly encourage you to use this process to effectively and efficiently communicate. Resist the temptation to immediately jump into writing the narrative. The up-front time that you devote to systematically following this process will pay significant dividends in terms of time, efficiency, and the quality of your finished document.

Information Packaging

Information packaging offers a variety of options to the technical writer. Technical information can be conveyed in several different ways: text (sentences and paragraphs), graphics, lists of information, schematics, information tables, and pictures. You, the writer, can assess all of these options and determine which packaging approach can be most effective for clearly and concisely transmitting the information.

PREPARING ENGINEERING REPORTS AND PRESENTATIONS

For the balance of this chapter, we'll focus our discussion on writing and presenting your co-op/intern report. This information will also be useful for written and oral presentations throughout your career.

ORGANIZATION MODEL

The basic organization model: everything you write—whether a one-page memo or a 15-page report—can apply the same organizational strategy. Your document must have a purpose and a message. It begins with a discussion of purpose. Why are you writing this? Why is it important? Why should readers be interested in this subject? This first section of your report provides background information, as appropriate, and perhaps references previous work or other relevant information. It creates context and a frame of reference for the reader.

The message is essentially the new information that results, for example, from your engineering project work or your research. For your intern report, the message will describe the work you did, the results you obtained, and the impact on (1) the company or organization and (2) you, personally, in terms of your technical growth and newly-gained experience. Where appropriate, a key element of the message may be your recommendations for future action, i.e., Where do we go next?

The bottom line for your intern efforts is a win-win experience for you and your employer—you win with a good grade for your excellent report and your employer wins by having your well documented work that may potentially enhance bottom-line business results. That last part is extremely important; when you conclude a communication, be sure to let the reader know any action that is required, i.e., "Where do we go next?"

Expanding upon this purpose-message structure, the basic organization model for your intern report is described.

STEP 1: Define Purpose

Introduction (Problem, Issue, Situation)

- ❖ Define the situation or issue, and why it is important.
- ❖ Describe the company/organization and its operations; create the context that describes the situation.
- ❖ Discuss important background, e.g., previous work in the same area.

Scope and Goal (Tasks)

- ❖ Describe the scope of the problem, in specific terms.
- ❖ Define specific objectives addressed and tasks performed.
- ❖ This provides a way to measure performance and serves to focus on your specific work.

STEP 2: Deliver the Message

Solution and Methods

- ❖ Describe the actions taken and methods used to address the issue.
- ❖ Discuss the value of this specific approach.
- ❖ Discuss the application of your engineering knowledge, including new methods.

Results and Conclusions

- ❖ Describe results achieved.
- ❖ Discuss extent to which results met project objectives.

- ❖ Discuss significance of the results to the project and the organization.
- ❖ Describe overall conclusions that can be drawn.

Recommendations

- ❖ Explain how the results should be implemented.
- ❖ Discuss need for continuing work in this area.

Your Growth Through the Internship Experience

- ❖ How did this experience benefit you technically?
- ❖ How did you grow in your people-related and business skills?
- ❖ Describe any expanded awareness of the engineering work environment.

DEFINING YOUR CONTENT

Information Needs: Level of Detail

Selecting the level of detail to include in your report is one of your most important and challenging tasks. To do this effectively, you need to thoroughly understand the purpose of your report and the background/experience of your readers, i.e., know your audience. The level of detail for information contained in your report can be visualized using a pyramid, as in Figure 25–3.

In most communication you'll probably be at level L-2 or L-3, wherein you'll describe the concept and include an appropriate amount of detail—suited to the reader's need to know. You need to provide the necessary information (evidence) to support your ideas, results, and conclusions. However, don't overburden your report with more detail than necessary. Ask yourself: What's important to the reader? What's not? Does it impact the specific project? If so, how? For example, if you were writing to a company manager, you wouldn't begin by giving background information about the company; the manager knows all that and you would be wasting his time and detracting

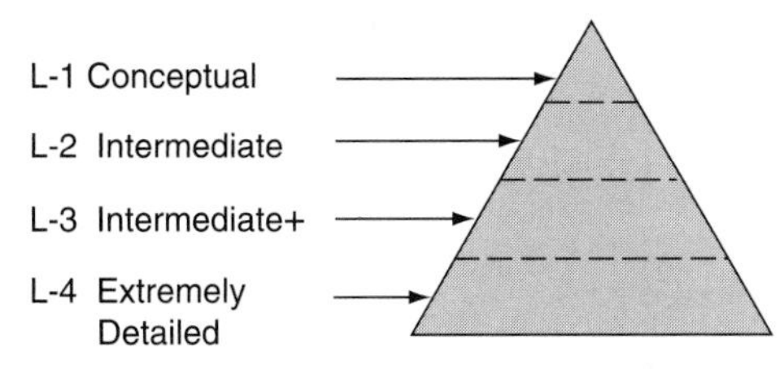

FIGURE 25–3 The Information Pyramid. L-1 is conceptual, with very little detailed information, appropriate for a senior manager. L-4 contains comprehensive detail, suitable for a peer engineer who was planning to continue your project work. Most reports are written at an intermediate level, depending on audience needs.

from the purpose of your communication. Get at the issue. Tell them what they don't know and give them enough background and context so that they can put your new information into the proper perspective. Remember: no redundant information; no irrelevant information. Be concise, but don't overdo it. Leaving out necessary information in the interest of conciseness isn't a good trade-off—you have to say it enough to get your point across.

OUTLINING—A KEY TOOL FOR ORGANIZING YOUR INFORMATION

Outlining is a technique that can help you get started on your report writing task. Before creating the content draft of your report, start outlining key information that you have gathered. This information would go under each heading of your engineering report, as indicated in the previous organization model. Outlining will start you on a path toward logical organization of your information that is geared to your readers' needs.

Purpose of Outlining

- Gather and organize your ideas for the report.
- List information points briefly.
- Identify any gaps in your information.

Benefits of Outlining

- It provides a basis for early review and planning.
- It allows you to define the scope, content, and organization of the report.
- It's an information checklist; do you have what you need?

Guidelines for Outlining

- Use the above model as a template.
- Express your thoughts briefly in sentence fragments.
- Define your information hierarchy. These will translate to sections and subsections of your report.
  - Major points
  - Supporting information
- Keep it simple and workable for you.

CREATING THE DRAFT

Now we can intelligently proceed to write a draft of our report. Keep in mind: Be consistent with your outline—include the correct information, presented in the correct sequence. Use headings to highlight sections and sub-sections so readers can easily follow your flow of information.

Sentences and paragraphs are the essential building blocks of your narrative report. They are complemented by other information packages such as graphs, schematics, tables, pictures, and so on. We'll discuss the integration of text and graphics later.

Effective Paragraphs

Effective paragraphs are separate thoughts organized to convey one major idea, i.e., a text package containing one main point and sufficient information (evidence) to establish the credibility of the main point.

Emphasize the major idea through good paragraph structure:

- ❖ Introduce the paragraph with a topic sentence that summarizes the main point.
- ❖ Follow the topic sentence with evidence that supports the main point.
- ❖ You may have a transition sentence that moves the reader to the next paragraph.

A suggested guideline on paragraph length:

- ❖ Shorter memos/letters—5–6 lines
- ❖ Longer reports—8–10 lines

The recommended paragraph structure is based on the concept of "main point first" with detailed information to follow. This main point first approach can be applied to sections and sub-sections of your report as well as to paragraphs. This writing style is of considerable benefit to the reader, who can efficiently read a document by keying on the main points and electing to read only the detailed information that may be of particular interest.

Guidelines for Sentences

- ❖ Each sentence should have one important point.
- ❖ Two or more important points in the same sentence may confuse the reader.
- ❖ Sentences of 15–20 words are the most effective in technical/business writing.
- ❖ Keep sentences short and easy to read.

Integrating Text and Graphics in a Document

You can use a combination of graphics and text in reports and proposals to help the reader to understand concepts and information. You can integrate graphics and text in a way that makes it easy for the reader to correlate the two. Judicious use of graphs, tables, schematics, charts, and sketches can be critical to the readers' thorough understanding of the technical material. Current text processing software offers us many options for effectively integrating text and graphics. Some practical guidelines are:

- ❖ Place graphics close to the related text in the document.
    - • Reference the graphics in the text (Figure Number, Table Number).
    - • Locate the numbered figure or table close to this text.

- Place the titles for figures, charts, schematics, and sketches beneath the particular graphic.

(text box)

GRAPH CHART SCHEMATIC SKETCH

Figure Number (Title or caption)

- Place the title for a table above the table.

(text box)

Table Number (Title or caption)

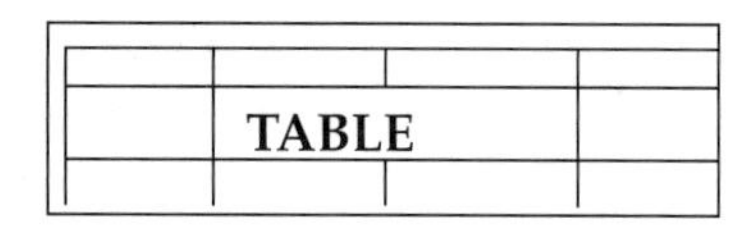

- Titles should be clear, concise, descriptive, and easily understood. You may select a range of fonts and types to differentiate your titles from the other text. However, be consistent throughout your document in your use of abbreviations and fonts.

Formatting

What is formatting? Formatting is structuring the general makeup of your report. It's the organization and arrangement of your material so that it can be most effectively understood by the reader. It gives your report a great first impression and invites the reader to read it because it looks good and clear. There are two types of formatting:

1. ***Page Design.*** This refers to how information is arranged on a typical page of your report. A key aspect of page format is the integration of text and graphics as previously discussed. Technical writing often contains text information in the form of vertical lists—a text formatting technique frequently applied in technical reports and written procedures. Other specific formatting techniques are outlined below.

   Summaries
   - Report: Include an executive summary, up-front, as part of any longer document.
   - Section: Consider starting each major report section with a brief summary of one or two paragraph that covers the main points.

   Headings
   - Use descriptive headings and subheadings that alert the reader to upcoming information. Note: headings with action verbs are effective.

- ❖ Your computer gives you many options for setting up headings and subheadings:
  - Capitalization of title words
  - A range of fonts
  - **Bold**, *Italics*, Underlining
  - At the margin, centered, indented
- ❖ Use your heading/subheading style consistently throughout your document.

Spacing

- ❖ Use white space effectively to avoid intimidating your reader. White space means wide enough margins and space between headings. You don't want to jam too many words onto a page from side to side and from top to bottom—it's hard to read.
- ❖ Use blocked (no indent), single-spaced paragraphs with an extra space between paragraphs.
- ❖ Make sure your page breaks are logical and easy to follow.
  - Don't carry one line of a paragraph to the next page.
  - Don't break up a listing to the next page.

Pagination

- ❖ Number all pages in your document, except for the title page and the cover letter.

Bullets—Use especially to:

- ❖ introduce several topics you'll discuss
- ❖ list a group of items, instead of using a long sentence or paragraph
- ❖ emphasize key points

2. ***Report Format.*** Typical organization:

- ❖ Cover letter (as needed)
- ❖ Title Page
- ❖ Table of Contents
- ❖ Executive summary
- ❖ Report sections (see basic model)
- ❖ References (as needed)
- ❖ Appendices (as needed)

The Executive Summary. The name says it all; a busy executive may only have time to look at a one-page summary of your report. Therefore, your objective with an executive summary is to make the reader aware of the key information without having to read your whole report. That means you will summarize the problem, your action or any alternatives, and major conclusions. A good executive summary of an important project may stimulate the executive to read your whole report. The executive summary is:

- ❖ An accurate synopsis of the report, addressing major elements in narrative style
- ❖ A stand-alone document
- ❖ A summary of principal points and critical supporting information
- ❖ Focused to appeal to busy executives

- ❖ A valuable part of major reports and proposals
- ❖ A good basis for a short verbal overview

In conclusion, the executive summary is not only a valuable, but often a critical component of your technical report.

Language, Readability, and Grammar

Have you ever read a report where you encountered, and were burdened by, such phrases as: "It is important to note" rather than "Note" or "engaged in the evaluation of" rather than "evaluated"? These glaring examples of "wordiness" often creep into our writing, almost without our realizing it.

In previous sections of this chapter, we discussed the value of planning and defining your audience, using a logical organization model, effective paragraph structure, and formatting and why they are a critical part of clear, understandable writing. However, an otherwise well-developed report may be severely compromised by poor language and grammar. This section briefly addresses this issue.

Four key shortcomings are responsible for the majority of the readability problems: excess words, inappropriate use of passive voice, lack of clarity, and poor grammar. Some flags for these problems are shown in Figure 25–4. It's your job, as the report writer, to identify these problems and edit to eliminate them. Let's examine each of the four in more detail.

1. ***Excess Words.*** Consider the following sentence:

   During the past two weeks, we have been wondering if you as yet have found yourself in a position to give us an indication as to whether or not you have been able to come to a decision on our offer. (41)

   The point is made more clearly and forcefully in the rewrite below:

   Have you decided on the offer we made you two weeks ago? (12)

2. ***Inappropriate Use of Passive Voice.*** A major contributor to wordiness is the excessive and inappropriate. Consider the passive voice sentence below:

   An analysis of the problem was conducted by the engineer.

   Note the weak sentence structure and tedious, roundabout expression of the idea. Rewriting, a clearer, more dynamic active voice structuring of the sentence would read:

   The engineer analyzed the problem.

   This version is direct and to the point, with appropriate emphasis on both the actor (who) and the action (what). Please note that there are many instances in technical writing where passive voice is most appropriate. Two examples come to mind:

   a. Instances where the action (i.e., what occurred) is more important than who performed the action (actor), or perhaps the actor is unknown.
   b. Instances where you, the writer, want to emphasize the object of the action rather than who performed the action. For example,

      Ten thousand field tests were conducted on the new prototype.

When editing for wordiness, look for the indicators or flags, namely: too many prepositions, e.g., **"to," "of," "in"**

Passive voice: forms of the verb **"to be"**

Phrases containing **"which"** and **"that"**

Overuse of noun forms of verbs, e.g., words that end in **"-tion," "-ment,"** or **"-ance"** (see passive voice example sentence)

FIGURE 25–4 Some Flags for Excess Words

In this case, passive voice construction emphasizes the large number of tests (action) conducted on the prototype (object).

3. ***Lack of Clarity.*** Some key problem areas:
   - Noun–pronoun association: the reader is unsure as to what a particular pronoun refers to.
   - Writing that is devoid of transition words, e.g., "however," "although," "therefore."
   - Incorrect or imprecise use of specific words or phrases.
   - Undefined jargon and acronyms.

   When reviewing your report for clarity, you must continually ask the question: "Will my report be clear to the reader, who is not nearly as knowledgeable on this subject as I am?" To review properly, you must assume the role of the typical reader.
4. ***Grammar and Punctuation.*** Some common grammatical errors:
   - Lack of agreement between subject and verb
   - Misplaced modifiers
   - Incorrect word usage and spelling
   - Erratic or incorrect punctuation, often involving use of commas

Addressing this area in more detail in this chapter would be redundant, considering the abundance of excellent references that already exist. Figure 25–5 includes a few references for your consideration.

Write Efficiently

"Engineer" your writing with a five-stage model:

Stage 1: Organize (outline) accurate, complete, and well-organized information

Stage 2: Create the Draft

- ❖ Focus on content; ensure clear purpose and message
- ❖ Develop graphics (tables, schematics, and so on)

Stage 3: Edit the Draft for language, refine graphics, prepare summary

Stage 4: Format the Report for easy reading and reference; prepare table of contents

Stage 5: Review, Finalize, and Issue give final review, proofread, revise, and issue

The OWL (Online Writing Lab), Purdue University
http://owl.english.purdue.edu/handouts/print/grammar

Word Usage in Scientific Writing
http://www.ag.iastate.edu/aginfo/checklist.php

Beer, D. and D. McMurray. *A Guide to Writing as an Engineer*, 2nd Edition. John Wiley & Sons, Inc., 2005.

Johnson-Sheehan, R. *Technical Communication Today*. Pearson Longman, 2005.

Perelman, L.C., J. Paradis, and E. Barrett. *The Mayfield Handbook of Technical and Scientific Writing*. Mountain View, California: Mayfield Publishing Company, 1998.

Strunk, W. Jr. and E.B. White. *Elements of Style*, 4th Edition. Needham Heights, Massachusetts: Allyn & Bacon, 2000.

FIGURE 25–5 Language and Grammar References

The five stages should be addressed in order; don't move to the next stage until you're reasonably well satisfied with the previous stage. This approach is the cornerstone of efficient writing. If applied faithfully, it can prevent major, time-consuming iterations in your writing.

It is most critical to develop a good content draft, concentrating on the logical presentation of your ideas and not worrying about editing at this stage. Attempting to do detailed editing at the same time as you are developing the content draft (i.e., the "write and polish" approach) disrupts your creative flow of ideas and results in time-consuming stops and starts. Detailed editing should be reserved for your completed content draft. See Figure 25–6 for a quick list of tips for good writing.

1. Think (plan, strategize) before you write.
2. Organize your information efficiently (user model in some form of outline).
3. Match writing style and level of detail to the audience and the situation.
4. Keep the focus on the reader.
5. Command interest and attention—"hook" your reader.
6. Convey a positive attitude.
7. Emphasize the active voice, use active verbs.
8. Create clear, concise sentences and paragraphs.
9. Use simple, straightforward language.
10. Streamline—eliminate redundancy and excess words.
11. Create an attractive, appealing format.
12. Use the five-stage approach, "Engineer Your Writing."
13. Write to inform, persuade, and get action.

FIGURE 25–6 Tips for Good Writing

ORAL PRESENTATION—PREPARE AND DELIVER

First, the bad news: Jerry Seinfeld says that most people fear public speaking more than death. He follows that by logically observing that, at the average funeral, most people would rather be in the coffin than giving the eulogy. He's joking, but for some people, that is dangerously close to the truth!

Now, the good news: public speaking can be taught, just like swimming. Neither public speaking nor swimming is done instinctively by humans. But once you know how to swim or give a speech, you'll find it to be a source of confidence, enjoyment, and even survival—whether in the water or managing your career.

PRESENTATIONS—SOME CONSIDERATIONS

Why Give an Oral Presentation?

At the university, it may be a requirement of your co-op program. On the job, it's an opportunity for you to discuss your project, present a new idea, or represent your company at a technical conference. An oral presentation offers some unique opportunities:

❖ *Immediate Information and Feedback.* This and the two following points mean that you can, essentially, give your written report on your feet to everyone who would read your report and get an immediate response. We will describe the use of visual aids and techniques for your oral presentation that can enhance the impact of your message.

❖ *Group Interaction.* After you make your presentation, your audience can ask any questions for clarification. It allows you to know that they understand the message. Group interaction can also create additional good ideas and allow the audience to be part owners of your message, especially when you accept feedback from them.

❖ *Understanding and Agreement (Buy-In).* Sometimes a good presentation can lead to an immediate decision to take action that you suggest in your message. If you're introducing a big idea for the first time, the decision would probably not be immediate, but your excellent presentation can go a long way toward a final decision that favors your proposal.

❖ *Development of Your Confidence and Communication Skills.* When a person is new to public speaking, it's natural for some nervousness and self-consciousness to exist. Then, with practice, any nervousness will absorb into enthusiasm to give your message to the audience. You'll forget about yourself and focus on the importance of conveying your ideas to the group. Still wary about public speaking? An old adage says "When there are butterflies in your stomach, make them fly in formation." That means you can turn nervous energy into passion to communicate your message. It works.

❖ *Career Growth Through Positive Visibility.* Recall Blair Webster's observation in Chapter 12, that "In a meeting your technical skills will be judged on your communication skills. If you are a good presenter, you will be judged to be a good engineer."

A convincing presentation has three elements, all of which must work together and in harmony:

1. **Verbal:** your words
2. **Vocal:** your tone of voice and enthusiasm
3. **Visual:** your body language, eye contact, sense of confidence, and urgency

Use all three "V"s to maximize your impact. Your words are important, but the vocal and visual impact of your talk may influence your audience's response as much as, or more than the words. If you are proposing an innovative, cost-saving idea to your management, but your body language and tone of voice do not convey your enthusiasm for or confidence in the idea, your otherwise positive message may not get through. In fact, more importance may be attributed to your non-verbal rather than your verbal message.

PREPARING YOUR PRESENTATION

Four Steps to Creating a Focused Presentation

1. *Know Your Audience*
   - Are they technical? Management? Know their make-up and background.
   - What are their interests? What are their expectations?
2. *Know Your Time Limit.* For example, your time limit may be 10 minutes total. Time limit is extremely important as you organize your presentation. Time passes more quickly than you realize while presenting, and you don't want to be only partially through your presentation when your time is up and you need to conclude. That means carefully organizing your presentation and practicing it to ensure that you can effectively give your message within the allotted time. Mark Twain once said, "I didn't have time to give a short speech, so I gave a long one." Similarly, when Winston Churchill was asked how much time he needs to prepare a speech, he responded, "If you want me to talk all day, I can start right now; if you want me to speak for five minutes, I need to do some serious preparation." Now, don't let those quotes scare you—they're just lighthearted realities that underscore the need to prepare for your presentation and satisfy all requirements and constraints, including time limit.

   For example, a seven-minute presentation would typically justify 8–10 visuals. In allocating your time, be sure to allow for questions and answers at the end of your presentation.
3. *Organize Your Presentation.* Make an outline of main points and detailed comments that fit within your time frame. Prepare notes that can be readily and easily referenced during your presentation.
4. *Create Effective Visual Aids.*

Presentation Format

- ❖ Title Page—title, to (name of audience), date
- ❖ Overview (tell them what you will tell them) with enthusiasm
- ❖ Content (tell them) with organized clarity:
  - Situation
  - Tasks and objectives
  - Action
  - Results and conclusions
  - Recommendations
- ❖ Summary and close (tell them what you told them) with enthusiasm and, if appropriate, a call for a decision or other specific action.

CREATING AND USING VISUALS

Effective Visuals

- ❖ Reinforce the speaker's message. They clarify and reinforce your message.
- ❖ Help the audience to absorb new information quickly.
- ❖ ***Common Characteristics of Effective Visuals.*** Simplified and, as a result, easy to read. Use the three Bs: big, bold, and basic. You don't want your visuals to be so complicated that they confuse the audience or detract from your presentation. Big, bold, and basic mean just that—relatively few points that are clearly visible and help you to communicate your message while helping your audience to understand it.

Types of Visuals

- ❖ Title Page
- ❖ Text visuals—these are well structured sentence fragments with headings and bullets so the audience can easily grasp the points you're making. In text visuals, make every word count. Make these readable or don't use them.
- ❖ Schematics
- ❖ Graphs and charts
- ❖ Tables

Working with Visuals

- ❖ You, the speaker are primary.
- ❖ Visuals are secondary and an aid to you.
- ❖ Introduce each visual and allow a few moments for the audience to see and absorb the information, while you also glance at the screen to remind yourself of key points you want to make. Then turn to your audience, resume eye contact, and emphasize the important points.
- ❖ Don't be afraid of a little silence when the audience is absorbing your visual information.
- ❖ Keep the lights up to connect with your audience and discourage sleeping.

DELIVERING YOUR PRESENTATION

Don't try to memorize your speech, and don't read it word for word. Instead, use notes and your visuals to guide you. Important techniques for effective delivery are discussed below.

- ***Eye Contact.*** Connect with your audience by looking at them. Some speakers never look up from their notes, while others might stare at the wall over the heads of their audience. As an effective speaker, you'll make it a point during your presentation to look in the eyes of various people throughout the room for about five seconds apiece. When you make eye contact, you're letting your audience know that you're talking to them all and welcoming them to accept your message. You'll occasionally be glancing at your notes and also the visuals to be sure you're on track, but meanwhile, eye contact is extremely important for you to personally and effectively connect with your audience.
- ***Posture.*** Besides eye contact, your body language is speaking, along with your words. As an effective speaker, you'll step before your audience with good posture and relaxed enthusiasm. Even before you begin to speak, your audience will have the impression that you want to be there and are about to share an important message with them. Your hands will be at your sides, perhaps with one hand holding notes or a PowerPoint remote. As you speak, you'll naturally gesture with one or two hands for emphasis as appropriate. Be natural. Not stiff. Not theatrical. Let your body and gestures flow with the speech and you'll be fine.
- ***Voice Level.*** You want everyone to hear you, but you don't want anyone to think that you're shouting at them. That means that you'll be speaking above normal conversational volume and appropriate for the size of the room and the audience. If using a microphone, you might begin by asking whether everyone can hear you.
- ***Voice Inflection.*** You have probably heard speakers make a presentation in a monotone voice that rivals a hypnotist in putting you to sleep. Did a presentation like that leave you with excitement? Or even a memory of the message? You are not that monotone speaker. You have an important message and your voice inflection will vary to emphasize certain points that you want to drive home. Voice inflection can include varying your pace and volume in a way that is natural for you and will make your presentation more interesting and effective.
- ***Handling Questions and Answers.*** At the end of your presentation you'll ask, "Are there any questions?" That shows that you're interested in knowing that the audience has absorbed your message and that you welcome any feedback. You may have to pause for a few seconds before the first person thinks of a question. That may invite other questions, which is good. Sometimes in a large audience, you may wish to plant a question to get discussion rolling. Questions mean that your audience is absorbing your message and thinking about how it can be applied. If the

DON'T	DO
• Read your presentation	• Dress appropriately; look professional
• Memorize your presentation	• Have a strong message; believe it
• Get interrupted	• Develop a strong opening and closing
• Exceed your time limit	• Know your stuff; establish credibility
	• Plan, prepare, practice

FIGURE 25–7 Presentation Don'ts and Dos

audience is large, you should repeat the question before you answer it. That ensures that everyone has heard the question and that you understand the question. Also, while repeating the question, you have a few valuable seconds to think of your response. Figure 25–7 provides valuable don'ts and dos for presentations.

DEVELOPING AND REFINING YOUR SKILLS

Your ability to give effective oral presentations can have a very positive impact on your engineering career success. Developing and refining your communication skills is a continuous journey. Over the years, your experience in giving presentations will allow you to improve your techniques and establish your reputation as an engineering leader. Meanwhile, you'll steadily increase your capability, confidence, and enjoyment in presenting your ideas before a group. Think about establishing an action plan for continuous improvement in this area. Talk with your manager and express interest in both feedback and the opportunity to make presentations whenever appropriate. That shows your initiative, leadership potential, and dedication to both to your career and your employer. Figure 25–8 shows additional tips for good presentations.

1. Have a strong message; believe it with conviction and passion.
2. Know your material; establish credibility.
3. Know your audience. Its expectations as well as your expectations.
4. Plan your presentation; take care of details and avoid surprises.
5. Organize your presentation logically.
6. Provide interesting, informative visuals to support your message.
7. Develop a strong opening and closing; know them cold.
8. Engage your audience; keep your cool.
9. Start on time; finish in time.
10. Practice! Practice! Practice!

FIGURE 25–8 Tips for Good Presentations

Chapter Summary: Takeoff Tips

- The abilities to write well and give a good presentation are crucial for your career.
- Those who communicate well are much more likely to be hired and promoted.
- The importance of communication in the workplace:
  - Engineers are required to communicate directly and frequently.
  - Communication skill is the necessary companion to technical knowledge.
  - Employers cite poor communication skills as a major competency gap.
- Know your communication objectives: To inform? To evaluate? To persuade?
- Good writing is easy to read and understand. It's coherent and lacks noise.
- Good writing requires that you know your subject and focus on the readers.
- Know the three Ps of technical writing: planning, process, and packaging.
- Know the organization model and steps for preparing your co-op/intern report.
- Engineer your writing with a five-stage model:
  1. Organize (outline) accurate, complete information.
  2. Create the draft.
     - Focus on content; ensure clear purpose and message.
     - Develop graphics (tables, schematics, and so on).
  3. Edit the draft for language, refine graphics, and prepare summary.
  4. Format the report for easy reading and reference; prepare table of contents.
  5. Review, proofread, finalize, and issue your document.
- The five stages should be addressed in order. Don't move to the next until you're satisfied.
- Review and act on Figure 25–6, Tips for Good Writing.
- As with effective writing, public speaking ability is a key for your success.
- Public speaking and swimming are not instinctive. Both skills can be taught.
- An oral presentation offers you some unique opportunities:
  - Group interaction—your audience can ask questions for clarification.
  - Understanding and agreement (buy-in)—you might get immediate action.
  - Development of your confidence and communication skills.
  - Career growth through positive visibility.
- Review and act on Figure 25–8, Tips for Good Presentations.

Exercises

1. Why should you have an objective to be an effective communicator?
2. How do you make your writing easy-to-read and understand?
3. What are the three Ps of technical writing?
4. What are the process steps in communication?
5. What are the organization model steps for your intern report?
6. Why is outlining a key tool for organizing your information?
7. What are the guidelines for creating a draft of your report?
8. Describe formatting.
9. What is the purpose of the executive summary?
10. What are the steps to creating a focused presentation?
11. What are important techniques for delivering your oral presentation?

CHAPTER 26

Transition to Industry

Now, in concluding this book, we are at the point of your career launch. So let's devote this last chapter to advice from fairly recent engineering graduates. We'll ask four engineers about their transition to industry along with tips they can give so you can really be ready for takeoff! Here are four young engineers with four different majors, to tell you their views about transition from campus to industry.

LINDSAY IVES

Lindsay Ives has a BS degree in Chemical Engineering. After her junior year, she accepted a co-op position with Praxair Inc. working in advanced control systems. After demonstrating excellent performance as a co-op student, Praxair offered her a full-time job when she graduated. She accepted this position and is now a process engineer with Praxair, doing process design of air separation plants. Here are Ms. Ives' thoughts on successful transition from campus to industry, starting with a co-op assignment.

Finding the Right Company

- *Location.* Do you have a location preference or restriction? Think about that as you accept a co-op or intern position, since your success as a student employee is likely to turn into a full-time offer.
- *Big vs. Small Company.* There are advantages to both large and small companies, which have been described in previous chapters. Target a large or small company based on your preference, skills, and career plan. In a small company you're more likely to be a generalist, with large companies giving more specialized assignments.
- *Work Environment.* Do you want to sit at a desk doing analysis and design or do you want to be out in the field doing hands-on work? In-house

design work will often lead to short-term field assignments; however, this may not be true if you begin a job in the field.

- *Travel.* Do you want to travel a lot, which could involve hands-on work, or do you have other commitments that would prevent travel? An internship or co-op assignment is a good time to ask how much travel is required for full-time employees in similar functions.
- *Pay, Benefits, and So On.* These are issues that come up later as you and the company assess your mutual value and correct placement within the organization. You should not bring these issues up during an interview.

When You Arrive at Your Job

- *Dress to Impress.* Don't be too casual, especially on your first day. Make any appropriate adjustments to your attire after you understand the job environment and what others are wearing. You wouldn't want to stick out like a sore thumb and make it obvious that you are an intern. Remember, dress for the job you want, not the job you have!
- *Elevator Speech.* Practice a 30-second elevator speech for use if a boss inquires about what you're doing. Say "hi" with a smile to people as they pass by. Use brief, friendly communication to let others know you're approachable and interested in the company.
- *Show Up on Time and Sober!* Punctuality is a big deal in industry and starting time may be much earlier than you're used to. It's a big transition from college, where you might be up until 2 am working on a project; now you're on an 8 am to 5 pm job. But there's no excuse for being late. Let them know you're there to work and make punctuality a habit.
- Your boss or someone in your group should show you around and introduce you to the people you'll be working with. Have your computer set up with certain programs you're going to need. Start working on your project during your first week. Often, busy companies are not ready for the intern, and will not have a well-organized task for you to work on right away. If this is the case, ask for applicable reading materials to give you a head-start on your assignment. Stay upbeat; don't be discouraged. Prepare to contribute as soon as possible.
- *Find a Mentor.* A mentor is a more senior person in the organization who can guide you through your co-op/internship and even as you start work after graduation. The mentor can guide you technically and in other ways, such as understanding the culture of the organization. The mentor may be your boss, or you can ask your boss to assign you one. Your mentor could be a key contact after your internship or co-op assignment ends and may help you get a job offer. Ms. Ives did stay in touch with her mentor, who gave her resume and a favorable recommendation to the right people to land the job she wanted in Praxair Process Engineering.

Tips While on the Job

- Your job may involve 25% technical and 75% communication skills. It's understood that as a co-op/intern you have not finished your engineering education, or even as a new hire you don't know the technology of the company. So it is very important to ask questions, and take notes so as to not ask the same question twice. It's also imperative to learn where to find technical information in your company so that you can start looking into resources on your own.
- Ask for work... and then more. Don't expect your boss to come to your office all the time and give you work, as you have been used to when given homework assignments. Your boss may be managing 10–20 other people, so he won't know all your deadlines or when you are ready for more work. So ask. Make sure you have enough meaningful work to do; regard that as your responsibility. That makes you look good—it shows that you take your job seriously and are ready to be accountable.
- Manage your time and meet your deadlines. As an intern or new hire, you might be given a job without a deadline; ask for a deadline and meet it. You're there to learn on the job; managing your time and meeting deadlines is an important part of any business.
- Meet with your boss on a regular basis. Don't let the only contact with your boss be the signing of your weekly time sheet. Set up regular meetings to let him know what you're working on and how you're doing. Ask for a performance review if you don't receive one, and always ask for ways to improve. That review could be informal and mid-way through your co-op/intern assignment. Then, at the end of your assignment, you should get another review and be able to show improvements to your boss.
- Find your niche. Get to know the company you're working for. Remember to ask about other areas of the company and learn what other job opportunities exist. You may find a position that interests you more than your internship, and that is okay. You might even consider asking your manager for project work in that area. Get to know what they need and what they're missing. Then decide where you can best fit—in order to work on a project and improve upon it toward the company's profits. Make yourself noteworthy. Make yourself a critical resource—companies do keep track of critical resources!

Landing a Job at Graduation

- Keep associated with your co-op/internship company. If it's a local company, try to remain working there after your co-op/internship period so you can continue to produce results and make yourself attractive as a prospective hire when you graduate.
- Be sure to put your co-op/internship experience on your resume. Describe the results you have achieved and be prepared to discuss your results during an interview. How you describe your experience is evidence of your accomplishments.

- ❖ Review your resume with your university's career services counselor and use your career office's services. They can help you successfully connect with employers. Smart students recognize this and fully utilize their career office resources. Sign up for campus interviews and research the company beforehand through literature and online so you can sell yourself by showing that you meet their needs during the interview. After researching the company, come up with a list of questions to bring to the interview. This will show you have come prepared to the interview and are interested in the company.
- ❖ Practice your interview skills and other soft skills. Often a technical interview is combined with a soft skill interview. Both are important and soft skills can be weighed as heavily as technical skills in making a hiring decision. Study Chapter 8 for typical technical and soft skill interview questions, so you can organize your ideas before the interview.
- ❖ At the interview: be prepared to talk about your co-op/internship assignment, other work assignments, and/or research experience. Explain your results, what you learned and why your organization benefited from your work. Give them evidence that you're technically sharp and can work well with people. Be yourself. Don't be afraid to show your true personality. Try to genuinely smile, be friendly, and look happy.
- ❖ Be persistent. Be sure to get an interview with the company where you had a co-op/internship assignment. Show interest. Find out when they will be making a hiring decision. Let them know your plans and suggest how you might fit their needs.
- ❖ Be sure there is a good fit. It's a two-way street. You need to be sold on the company just as importantly as the company needs to be sold on you. You want to find a job that you're happy with!

NATE MAERZ

Nate Maerz graduated with a BS degree in Industrial Engineering and a Six Sigma Black Belt with co-op experience at API Basco and Niagara Thermal Products. After graduation, Mr. Maerz worked at Polymer Conversions Corp, and now he's an industrial engineer with Mod Pac Corporation. Here are Mr. Maerz's views about transitioning to industry.

Getting Hired

- ❖ As you write your resume and prepare to job hunt, consider all that you have to offer. Certainly your coursework is important, but also important are the extra skills you can bring to an employer, including job experience, leadership in school clubs, volunteer activity results and credentials such as Six Sigma Black Belt and Lean. Six Sigma and Lean are important concepts and buzz words in industry. Your extra qualifications in areas such as these will set you apart and inspire an employer's interest.

- Post your resume on your university's career services Web site and regularly check for postings. Use the Internet. Check newspapers. Use your network of friends and relatives to get leads. Send your resume wherever you want to be considered.
- Go on lots of interviews. That gives you interviewing experience and gives you the chance to compare opportunities at different companies. Could you see yourself working at these companies? Which is best for you? Where will you be happiest and make the most impact?
- During interviews, be yourself, but be professional and discuss how you can be valuable to that employer. The company is deciding on you, but you are also deciding on the company. You don't want to work in a place where you can't function well and make a good impact.

What Employers Are Looking For

- An employer might advertise a need for someone with 2–5 years experience. You don't have that coming right out of college. But if you have the right technical skills and interest, go ahead and apply. You might be the best candidate, despite the experience qualification stated in an ad.
- Employers are looking for someone who wants to work for them. That means someone who is competent, willing to be trained and learn new things, including outside your discipline. Employers are looking for someone who can perform and who they can count on. They want the confidence to give you an important project and know that you will meet a deadline.

Stepping into Industry—It's an Adjustment

- No more sleeping until noon, except on weekends.
- Late nights means you're at work working late; you're not out partying.
- Going to work is not optional, like going to elective classes.
- You'll dress appropriately for your job, no sweats.
- You may have homework if you can't get it done during the day.

School to Industry

- Your engineering coursework prepares you well for industry's technical needs.
- Missing from a standard engineering curriculum is exposure to the soft skills, the people skills that are important and that you'll need in industry. You'll learn communication skills, including explaining project-related technical concepts to non-engineers, such as financial people, in layman terms that can be understood.
- You'll discover people who you can go to for answers and who'll work well with you. Some will join you in solving problems while others may not be as helpful and may even hinder you. Learn this, and if you have a choice, work with those who can enhance your productivity and results. You may not have a choice of teammates, so in any case be friendly,

professional, and focused on project goals, not personality differences. You may be leading a team, in which case your leadership skills will be very important.

- There are deadlines on projects, and you will quickly learn the importance of meeting those deadlines.

In the Workplace

- Be aware of the company's culture—its rules and ways of operating. Each company is different. Some companies have a 9 to 5 job and everyone leaves at the same time. Other companies have the expectation that you'll work extra hours, and you might have to come in at midnight to train the third shift on something.
- Be punctual. Arrive at work on time and don't be the reason for a meeting to start late.
- Be noticed. Get to know your boss and co-workers. Go to lunch with them. You want to be noticed, but also, be discrete: you don't want make it obvious that you're campaigning for a promotion. Be noticed positively as a team player.
- Have an "elevator speech" ready: that's a 30–60 second summary of what you're working on in case you're on an elevator or in a limited time situation with an executive who asks what you're doing. Those are moments to summarize your project and results. Take advantage of those opportunities as a big step in being noticed.

Some Challenges You May Face

- Get a clear definition of your project. In your co-op/internship or assignment as a new hire, there should be a clear definition of what your project is and what you should be doing. After all, you're being hired because they need you. In reality, your boss may have recognized a problem or opportunity that needs attention; however, the details of the project might not yet be defined. You don't want to be hounding your boss to the point of irritation, so you may need to develop a project definition draft and then propose an action plan to your boss for approval. The good news: in such a case, you have just gotten experience defining a project yourself and now you have a clear and approved project definition you can go with. That's a project management career growth step for you—recognize it as such!
- Define the project so that it works for you. Don't volunteer for more than you can handle. Don't try to take on a big project with a deadline that's unreasonable. If you are on such a project, you might scale it down and suggest it be divided into segments so that you can feasibly finish the first part on schedule, then negotiate a reasonable action plan, including a deadline for the second part.
- As an industrial engineer, your work is very project-based. Often there are several projects going on at once. Prepare to handle these multiple

projects, including planning and progress meetings. Expect that there'll be project teams and that you'll learn to work effectively with the people on the teams. If you're a project leader, you'll want to ensure that everyone on the team is doing their part—so you can rely on them to get their part of the job done correctly and on schedule. If you're a team leader, you might be able to choose the people for the team; in that case, you might be able to select team members you can work best with and who'll produce the best results.

- Stay busy. Find things to work on that'll be helpful and that your boss hasn't given you. Present your proposal to the boss, who should be impressed with your initiative and that he can count on you as a valuable employee.

Go Out on the Plant Floor

One of the most important things Mr. Maerz learned—go out on the plant floor, meet the employees and learn from them!

- Machine operators, maintenance technicians, material handlers, and supervisors—they're out there every day performing processes that you're supposed to be working on. They have the greatest knowledge and the best experience to teach you what they're doing. Get to know them. Listen to them. Respect them. Learn and gain respect.
- Plant workers can tell you problems that affect productivity. They'll tell you about an inefficient process that they have been asked to perform. They can also make suggestions for improvement on a broader scale. This information can lead you to valuable projects for improvement. It might be an ergonomic study or an evaluation of plant layout. The workers can also lead you to solutions for the project. The people in the plant are the greatest resource you have. So, get out on the floor and talk with those people! Your results and your career will move ahead that much faster.

JASON LUPIENSKI

Jason Lupienski has BS and MS degrees in Electrical Engineering and co-op experience with Motorola, Carlton Technologies, and Sierra Research. He has had full-time experience with Syracuse Research as a systems engineer and General Motors Powertrain as a controls engineer. Now Mr. Lupienski is a project engineer with General Motors fuel cell activity, working on the next generation of automobiles, in this case hydrogen fuel cell cars. Based on his experience, Mr. Lupienski has some suggestions for anyone going from campus to an industrial career.

Plan for the Job You Really Want

- For the first two years, be a technical sponge. Try to get the most technical knowledge possible. Those first two years are critical, since afterward you'll be broadening in many different directions. The first two years will be your technical foundation, applying your engineering coursework while continuously learning more about different products and processes.

- Become rounded and versatile. Just because you are an electrical engineer doesn't mean you won't be asked to do mechanical functions. In a total team environment, everyone is a systems engineer putting a lot together. It could be mechanical, electrical, fuses, structures, and so on. A systems engineer is well-rounded and can function as a quality engineer, mechanical engineer, electrical engineer—whatever is needed. Mr. Lupienski, the MS Electrical Engineer, does more mechanical than electrical work and does more software engineering than anything.
- Continue your education. Your BS degree is your basic entry to engineering, but it doesn't stop there. Mr. Lupienski suggests you go on for your master's degree or MBA, since the world of technology and business is becoming increasingly complex and competitive. Continue your education with certificates and specialized training. For example, become a Six Sigma Black Belt.
- Decide where you want to go. What boxes on a list do you need to check in order qualify for advancement? Take the steps and learn the things needed to qualify you for your goals. Your company may have plans for you, but only you can decide what you want to do and where you want to go, including how high or wide.
- Don't get pigeonholed. That means don't allow yourself to become more specialized than you want to be, especially early in your career. If you're the best designer in the group and seemingly indispensable your boss might not want to let you go, even though you should be getting more experience in other areas. Know that you're good and that you have many opportunities. Where do you want to go? What do you want to do? You're in charge of your career.

Balance Your Life

- It's not all work—make sure you go out and have fun. Work hard and play hard. When you arrive on the job, get to know some people you can go out and socialize with.
- It's important, especially in the long term, to balance your professional and personal life. Dedicating yourself to your job, but also to your family and other activities make a well rounded life, while allowing you to be healthiest and happiest.

Have Passion for Everything You Do

- Even if you get stuck with a lousy project, have passion for doing it well. Promote the importance of that project. Let it be known that you'll make money for the company and that a patentable idea may result.
- Your passion will be noticed and will make a positive impact that'll enhance your career and home life. Whether you're an intern/co-op student or a new hire, make sure you show passion and let the company know you

really want to help them succeed. Mr. Lupienski concludes: "show that passion and you'll get promoted so fast it will be unbelievable."

BEN KUJAWINSKI

Ben Kujawinski has a BS degree in Civil Engineering. He had an intern assignment at Taylor Devices, did a great job, and was hired there upon graduation. Mr. Kujawinski is now Production Manager at Taylor Devices. Here are Mr. Kujawinski's thoughts about engineering and transition.

Why Study Engineering?

- ❖ Engineering is a core skill sought by every company, technical or otherwise, everywhere in the world.
- ❖ The foundation of engineering is solving problems for good and for profit.
- ❖ An engineering degree is a unique start to an industrial or academic career. An engineering degree also opens the doors to many other professions such as medicine, law, or finance.

Getting Through Engineering School

You don't need to be told that engineering study is difficult. Here are some ideas about "how to get through it" to graduation, especially if you're struggling:

- ❖ Don't be too hard on yourself. Recognize that this is a challenging curriculum and don't beat yourself up when the going gets tough.
- ❖ Get involved. There are engineering clubs, projects, and activities that will open doors and provide opportunities to meet helpful people. Involvement can reveal hidden possibilities, showing the fun and applied side of engineering. Enjoy. As an undergraduate, Mr. Kujawinski was editor of the Engineers' Angle student newspaper and coordinated Engineers' Week celebration events.
- ❖ Develop support. That can include family, friends, professors, and teaching assistants.
- ❖ Use your university's resources, including your career services office. They're there to help you while you're a student and during your career.

Your Transition from School—A Unique Experience

- ❖ You, like everyone, will have a unique transition—based on your skills, interests, goals, and options from which you can choose.
- ❖ Keep an open mind to the many opportunities that are out there. That can include work within your engineering major, outside your major, or even outside of engineering. Mr. Kujawinski has friends with an engineering degree who stayed with engineering and others who are now in law, medicine, management, and other fields where an engineering degree provides an excellent educational foundation.

The Good	The Challenging
• Broad exposure to the business	• There is a lot to learn
• Advancement can be rapid	• Stress can be high
• Family atmosphere	• Close ties can cause conflict
• High compensation potential	• May not get a high initial offer

FIGURE 26–1 Aspects of Engineering Employment in a Small Company

Going to Work in a Small Company

Mr. Kujawinski summarizes the good and the challenging aspects of engineering employment in a small company in Figure 26–1.

In all, Mr. Kujawinski is very happy with his work at Taylor Devices, a small company. He recognizes that there are tradeoffs in any situation, including a large or small company. Be aware that in many companies technology will be only part of what makes business run. Accounting and other non-engineering functions are also important and play a big role in making the overall business come together and click.

In General

❖ Be prepared for varied personalities and be flexible in dealing with them.

❖ Some days will be not so good. Take the good with the bad, and recognize that the bad days are not the end of the world. Usually tomorrow is better than the bad day.

❖ Develop relationships. How you treat people will be remembered more than how often you were right.

Transition Keys

❖ Keep an open mind to the opportunities before you. As indicated, you're best able to decide which opportunity is best suited to your profile of abilities and interests.

❖ Make good first impressions. You won't have a second chance for a first impression.

❖ Learn and grow from mistakes. Everyone makes mistakes. Use them for growth, but avoid them in the future.

❖ Use the Golden Rule—treat others as you want to be treated.

❖ Avoid debt like the plague, especially credit card debt that has high interest rates. You won't make a ton of money after you graduate and living on your own will cost more than you think.

❖ Stay, or become, well rounded and balance your lifestyle to avoid burnout. Exercise. Pursue sports and hobbies. Try something new. You'll be healthier and happier, especially in the long run.

PANEL DISCUSSION

The following are questions posed by engineering students with our panel of engineers' responses:

1. **What should I do when I first get onto the job?**

   A: First impressions are important. Dress neatly and arrive on time.

   A: Be polite, friendly, and interested in learning—be motivated to contribute.

2. **Can I ask too many questions?**

   A: Perhaps, if you overdo it. Don't ask repeat questions; avoid this by writing down the answers the first time.

   A: Ask enough so you understand, but don't go overboard.

   A: Questions are expected, but don't take too much of your supervisor's time. Collect your thoughts and learn from experience, especially if you've been given an explanation of your question previously.

3. **Do you refer to your engineering textbooks often on the job?**

   A: It depends on your job. I refer to my thermodynamics text and some others, but the main thing is to know where to find an answer. You can go to someone for help, but you need your basic technical foundation books.

   A: Keep your textbooks—you never know when a technical question related to your job will come up, and you can't fake a technical answer. Never fudge your answers or data—that could cause serious safety or financial consequences.

4. **Does the fun stop when you get a job?**

   A: No, the fun doesn't stop. It's a different kind of fun from school. It can be more fun because you're applying your coursework, making a product, and making money.

   A: Balance your work and your personal life. Pay attention to both. That helps to keep you happy and successful in each area.

   A: As an intern, co-op, or new hire, be aware and don't get carried away with fun—don't come to work after a party without sleep.

   A: You can find people at work who have common interests that will lead to good and lasting friendships at work and outside of work. It's not uncommon to find your future spouse among your fellow employees.

5. **Do you have any advice about paying off loans?**

   A: You have to repay your student loans, so budget for it.

   A: Avoid credit card debt. Big credit card debt will rip the fun out of your life. Carry zero credit card debt by always paying promptly.

   A: Beyond repayment of student loans, an intelligent loan is a mortgage on well selected real estate at a low interest rate.

A: Get educated about your finances and personal financial planning. Some financial advisors say you should save and invest as much as 20% of your income so that it can grow over the years through compound interest. That includes tax deferred, diversified investments. Don't put all your eggs in one basket.

6. What's the most effective approach in getting a job?

A: Start early. Get your resume ready and pursue all the employment opportunities that meet your objectives. That goes for internships, co-ops, and graduating engineers.

A: Use your career office and all the other sources to connect you with possible employers. That includes friends, relatives, professors, LinkedIn contacts, and anyone they can refer you to.

A: Who are the employers that match your skills and interests? Research them on the Web and through your career office so you can be prepared for an interview. Know what they do and how you can be valuable to them.

A: Don't give up. Persist with the confidence that your search will pay off with the right job. In a job search, you'll get out of it what you put into it. It's definitely worth the effort of finding a good job.

Chapter Summary: Takeoff Tips

❖ *Starting a Co-Op/Internship*

- Find the right employer based on your personal and professional preferences.
- When you arrive at your job, follow the advice in Chapter 10, including:
  1. Dress to impress. Don't be too casual, especially on your first day.
  2. Be punctual, friendly, and prepared to contribute as soon as possible.
  3. Find a mentor in addition to your boss.
- Your job may involve 25% technical and 75% communication skills.
- Ask questions and take notes so you don't ask the same question twice.
- Learn where to find technical information so that you can do so on your own.
- Ask for work . . . and then more. Ensure that you have enough meaningful work.
- Manage your time and meet your deadlines.
- Meet with your boss on a regular basis. Communicate how you're doing.
- Find your niche with your employer. What opportunities suit you best?
- Get to know what your employer needs. Where and how can you best help them?
- Make yourself a critical resource. Companies keep track of critical resources.

- ***Following a Co-Op/Internship Assignment.*** Keep associated with that employer:
  - You might continue working part time and produce impressive results.
  - Make yourself attractive as a prospective hire when you graduate.
  - You might stay in touch with your mentor, who may help you get hired.
  - Be persistent and show interest. Be sure to get an interview.
  - Find out when they will be making a hiring decision. Convey your plans.
- ***Use Your Career Center Services.*** Take advantage of your career center services to pursue your options:
  - Be sure to put your co-op/internship results on your resume.
  - Post your resume in the career center Web site and check for postings.
  - Practice your interview skills and other soft skills.
  - Go on lots of interviews.
- Network with friends, relatives, LinkedIn, Web, and so on for leads. Pursue those leads.
- At interviews, you'll talk about your co-op/internship/research assignments.
  - Explain your results and why the organization benefited from you.
  - Give evidence that you're technically sharp and have people skills.
- Be sure there's a good fit. You want a job that you're happy with.
- Employers are looking for someone with desire, competence, and reliability.
- Stepping into industry is an adjustment: early punctuality, dress, commitment.
- Your engineering coursework prepares you well technically.
- Soft skills are needed. Soft skills are missing from most engineering curricula.
- Projects have deadlines. You'll learn the importance of meeting deadlines.
- Be aware of your employer's culture. Fit in. Each organization is different.
- Be noticed positively as a team player. Get to know your boss and co-workers.
- Have an elevator speech ready for a progress report to a boss.
- Prepare to handle multiple projects, but don't volunteer for more than you can do.
- Meet your fellow workers and learn from them. Respect them and gain respect.
- To start, be a technical sponge. Get the most technical knowledge possible.
- Become well-rounded and versatile.
- Continue your education for an MS, MBA, or PhD.
- Decide on your career goals. Take steps needed to qualify you for your goals.
- Early in your career, don't become more specialized than you want to be.
- Balance your life. It's not all work; make sure you go out and have fun too.
- Have passion for everything you do; even a lousy project—do it well.
- Your passion will be noticed and will make a positive impact on your career.

- ❖ Engineering is a core skill sought by every employer in the world.
- ❖ The foundation of engineering is solving problems for good and for profit.
- ❖ An engineering degree is a unique start for an industrial or academic career.
- ❖ An engineering degree also opens the doors to medicine, law, and other professions.
- ❖ Getting through engineering school is difficult. Don't beat yourself up.
- ❖ As an undergrad, get involved, develop support, and use resources to help.
- ❖ Be realistic: there are good and challenging aspects associated with any employer.
- ❖ Keep your engineering textbooks; you can't fake a technical answer.
- ❖ You have to repay your student loans, so budget for it.
- ❖ Avoid credit card debt.
- ❖ Get educated about your finances and do personal financial planning.

Exercises

1. As our final chapter for your engineering career launch, we have asked four young engineers with four different majors to tell you their views about transition from campus to industry. Please describe what you learned from the following people:
   a. Lindsay Ives
   b. Nate Maerz
   c. Jason Lupienski
   d. Ben Kujawinski
2. State three or more things that you learned in the panel discussion.
3. Do you feel more prepared for your career after reading this book? Why?
4. Make a life-plan for your professional and personal future.

Final Thoughts

You're holding an engineering-success handbook geared to give you a jump-start to your career. Professionals agree that everything we have presented is important to your engineering future. Your engineering coursework is vital, but application of the contents of this book will make a big supplemental difference in your career. If you have chosen to learn and use these subjects, you'll be far ahead of your contemporaries as you graduate with a good job and launch your career.

Wishing you the best of success and happiness throughout your life!

RECOMMENDED READING FOR SUCCESS

These books will supplement *Ready for Takeoff!* and provide you with a valuable library for your career success. The books on this list have been carefully chosen for value, relevance, and reasonable cost.

Bennis, Warren. *On Becoming a Leader.* Basic Books, 2009.

Bolles, Richard. *What Color Is Your Parachute?* Ten Speed Press, 2009, (job hunting).

Carnegie, Dale. *How to Win Friends and Influence People.* New York: Simon and Schuster, 1981.

Covey, Stephen. *The Seven Habits of Highly Effective People.* New York: Free Press, 2004.

Friedman, Thomas. *The World Is Flat.* Picador, 2007, (business globalization).

Glickman, Rosalene. *Optimal Thinking.* John Wiley & Sons, 2002, (how to be your best self).

Hoff, Ron. *I Can See You Naked.* Andrews McMeel Publishing, 1992, (fearless public speaking).

Kriegel, Robert. *If It Ain't Broke . . . Break It.* Business Plus, 1992, (staying successful).

McGraw, Phillip. *Life Strategies: Doing What Works, Doing What Matters.* Hyperion, 2000.

Morrison, Michael. *First Time Manager.* Kogan Page, 1999.

Stanley, Thomas, and William Dank. *The Millionaire Next Door.* Pocket, 1998, (financial success).

Strunk,William. *Elements of Style.* Waking Lion Press, 2009, (effective writing).

Vickey, Jesse. *Life After School Explained.* Cap & Compass, 2002.

White, William J. *From Day One.* Prentice Hall, 2010, (career success).

Yate, Martin. *Knock 'em Dead 2010: The Ultimate Job Search Guide.* Adams Media, 2009, (interviewing).

INDEX

M

S

W

X

Y